T0271000

Principles of Organic Farming

About the Authors

Dr E. Somasundaram, is the Professor and Head of the Department of Sustainable Organic Agriculture, Directorate of Crop Management, and PRO of Tamil Nadu Agricultural University, Coimbatore, involved in education, teaching, research and extension activities on organic agriculture in the University. He has completed M.Sc. (Ag.) with University rank and won S.R.P.P. Chettiar Gold Medal and Shield, A.C Mudaliar and Rao Saheb Murugesa Mudaliar awards for the best M.Sc. (Ag.). Student of the University. He has also won PPIC Gold Medal for the best M.Sc. (Ag.) Thesis on resource conservation and recycling for rice yield maximisation and IOB Gold medal for the best M.Sc. (Ag.) Student in Agronomy. He is the recipient of Prof. Subramanian and Dr.K.K.Subbiah awards for the best Ph.D., student in Agronomy. Recently he received the best Professor in agribusiness management during 24th B School awards at Mumlai. He visited several countries for presenting papers. He published many books besides having 50 research publications and many popular articles to his credit. He continues to guide M.Sc. and Ph.D. students.

Dr. D. Udhaya Nandhini, Post Doctoral Fellow, Centre of Excellence in Sustaining Soil Health, ADAC & RI, Tamil Nadu Agricultural University, Trichy. She has completed her M.Sc. in Agronomy in the same situation with reference to cropping system in pigeonpea. She was awarded with Student Senior Research Fellowship and worked on stress physiology to study the effect of Lipo Chito Oligosaccharides (LCO) compounds on maize and blackgram under stress conditions. She was the recipient of Adarsh Vidya Saraswati Puraskar Award. She authored 2 books besides having 30 research publications and several popular articles to her credit.

Dr. M. Meyyappan obtained his B.Sc. (Ag.), M.Sc. (Ag.) Agronomy and Ph.D in Agronomy from Annamalai University. He secured first rank in M.Sc. (Ag.) Agronomy. He is working as Assistant Professor of Agronomy in Annamalai University from 2000 onwards. He is teaching various subjects for Diploma in Agriculture, B,Sc. (Agri and Horti.), M.Sc. (Ag.) Agronomy and Ph,D Agronomy students. He participated and presented papers in eleven international conferences and twenty five National seminars. He published twenty six research papers in various journals and conference proceedings. He guided six M.Sc. (Ag.) Agronomy students. He authored individually for two books, Basic Text Book on Forestry and Notes on Water Management during 2012 and 2013 respectively.

Principles of Organic Farming

E. Somasundaram
D. Udhaya Nandhini
M. Meyyappan

CRC Press is an imprint of the
Taylor & Francis Group, an **informa** business

First published 2022
by CRC Press
4 Park Square, Milton Park, Abingdon, Oxon, OX14 4RN

and by CRC Press
6000 Broken Sound Parkway NW, Suite 300, Boca Raton, FL 33487-2742

CRC Press is an imprint of Taylor & Francis Group, an Informa business

British Library Cataloguing-in-Publication Data
A catalogue record for this book is available from the British Library

Library of Congress Cataloging-in-Publication Data
A catalog record has been requested

ISBN: 978-1-032-19784-5 (hbk)
ISBN: 978-1-003-26084-4 (ebk)

DOI: 10.1201/9781003260844

TAMIL NADU AGRICULTURAL UNIVERSITY

Dr. K. Ramasamy, Ph.D.
Vice-Chancellor

Coimbatore - 641 003
Tamil Nadu, India

Foreword

Global interest has been increased on organic farming as it has diversified benefits to the society and environment. The rapid growth of organic farming at global scale started during the implemented and movements were promoted by both governmental and non-governmental organizations. This led to rapid development of organic farming with co-ordinate and rational approach. By the end of 2017, about 50.9 million hectares of organic agricultural land including in-conversion areas reported around 179 countries in the world showing positive growing trend as compared to yesteryears.

The information on organic agriculture is scattered while the interest on subject is increasing linearly day by day. Therefore this effort with the aim to fill the gap of need on various aspects of organic farming and its publication is quite appropriate and timely. The present book title "Principles of Organic Farming" provides solid background information that is accessible for those who have a strong interest in organic agricultural education, research and development.

I hope that the book will cater the needs of undergraduates and postgraduates studying organic agriculture and sustainable crop production. The interpretation of subject matter in each chapter of this book has also been made carefully and systematically by the authors giving more emphasis on the requirements at the agricultural university level education in the country.

I have no doubt that this text book will go through many editions, being reworked and updated to become a standard reference for those pursuing courses on organic farming in the country with the days to come.

The efforts by Dr. E. Somasundaram, Dr. Udhaya Nandhin and Dr. M. Meyyappan to develop the theme for each chapter and transform into a valuable publication deserves appreciation.

I am sure that this textbook would be of great value for the students in particular and will be useful to the teaching community at large.

Place: Coimkbatore

(K. Ramasamy)

Dated 17-10-2018

Preface

The quantitative sufficiency of food in our country has led into think the maintenance of soil health and crop husbandry techniques, which maintain the nature's balance. Organic agriculture (OA) can be seen as pioneering efforts to create sustainable development based on other principles than mainstream agriculture. Practices of organic farming are resilient and becoming increasingly important due to pressing needs to protect the air, soils, and water; to improve socioeconomic conditions of farmers, farm workers, and rural communities; and to provide healthy, safe, and nutritious horticultural products to a rapidly increasing world population.

Organic agriculture has always been India's inherent advantage and strength. The shift in the global consumption patterns, health awareness among the consumers and the increasing significance of sustainability is now putting Indian organic products to the forefront both internationally as well as in the domestic market.

Principles of Organic Farming is a practical oriented text to organic crop management that provides background information as well as details of ecology-improving practices. This book is meant to give the reader a holistic appreciation of the importance of organic farming and to suggest ecologically sound practices that help to develop and maintain sustainable agriculture. This book represents a current look at what we know about organic farming practices and systems, primarily from the Indian perspectives.

This book is intended as a professional basic textbook for undergraduate level students and will specifically meet out the requirement of the students of **Organic Farming** being taught at many of the universities. In addition, the purpose of this work is to spread the basic concepts of organic farming in order to; guide the production systems towards a sustainable agriculture and ecologically safe, obtain harmless products of higher quality, contribute to food security, generating income through the access to markets and improve working conditions of farmers and their neighbourhoods.

This book provides attention of one and all concerned to promote organic farming as a measure to provide the elutes to posterity and to save our farm land that inherited from forefathers from being degraded and made in to wastelands through our excessive interventions.

The authors thank Dr. K. Ramasamy, then Vice-Chancellor, Tamil Nadu Agricultural University for providing necessary guidelines for bringing out this book.

The authors thank their family members for their help, physical and moral support for their sincere efforts in bringing out this book in time.

Authors

Contents

Chapter 12: Post Harvest Management of Organic Produces 307

Chapter 1
Exordium

Organic farming system in India is not new and is being followed from ancient time. It is a method of farming system which primarily aimed at cultivating the land and raising crops in such a way, as to keep the soil alive and in good health by use of organic wastes (crop, animal and farm wastes, aquatic wastes) and other biological materials along with beneficial microbes (biofertilizers) to release nutrients to crops for sustainable production.

Presently the farming situation urges need to develop farming techniques, which are sustainable from environmental, production, and socio-economic points of view. Modern agricultural production throughout the world does not appear to be sustainable in the long run.

Sustainable agricultural development "is the management and conservation of the natural resource base and the orientation of technological and institutional change in such a manner to assure the attainment and continued satisfaction of human needs for the present and future generations". Such sustainable development in the agriculture, forestry and fishery sectors, conserves land, water, plant and animal genetic resources, is environmentally non-degrading, technically appropriate, economically viable and socially acceptable. Consequently these concerns imparted a way to organic farming. It is the need of the day to understand the prospects and problems of organic farming to launch a successful and flawless organic production programme in the farm environment.

There are many explanations and definitions for organic agriculture but all converge to state that it is a system that relies on ecosystem management rather than external agricultural inputs. It is a system that begins to consider potential environmental and social impacts by eliminating the use of synthetic inputs, such as fertilizers and pesticides, veterinary drugs, genetically modified seeds and breeds, immunity booster preservatives, additives and irradiation. These are replaced with site-specific management practices that maintain and increase long-term soil fertility and prevent pest and diseases.

Definition

Organic farming "is a production system which avoids or largelyexcludes the use of synthetically compounded fertilizers, pesticides, growth regulators, and livestock feed additives. To the maximum extent feasible, organic agriculture systems rely upon crop rotations, crop residues, animal manures, legumes, green manures, off-farm organic wastes, mechanical cultivation, mineral bearing rocks, and aspects of biological pest control to maintain soil productivity, tilth, to supply plant nutrients, and to control insects, weeds and other pests". (USDA,1980).

Organic farming is a holistic production management system which promotes and enhances agro-ecosystem health, including biodiversity, biological cycles, and soil biological activity. It emphasizes, the use of management practices in preference to the use of off –farm inputs, taking into account that regional conditions require locally adapted systems. This is accomplished by using, where possible, agronomic, biological and mechanical methods, as opposed to using synthetic materials, to fulfil any specific function within the system (FAO, 1999).

IFOAM describes "Organic Agriculture is a production system that sustains the health of soils, ecosystems and people. It relies on ecological processes, biodiversity and cycles adapted to local conditions, rather than the use of inputs with adverse effects. Organic Agriculture combines tradition, innovation and science to benefit the shared environment and promote fair relationships and a good **quality of life** for all involved."

The concept of the soil as a living system which must be"fed" in a way that does not restrict the activities of beneficialorganisms necessary for recycling nutrients and producing humus is central to this definition.

In philosophical terms organic farming means "farming in spirits of organic relationship. In this system everything is connected with everything else. Since organic farming means placing farming on integral relationship, we should be well aware about the relationship between the soil, water and plants, between soil-soil microbes and waste products, between the vegetable kingdom and the animal kingdom of which the apex animal is the human being, between agriculture and forestry, between soil, water and atmosphere etc. It is the totality of these relationships that is the bed rock of organic farming.

Concepts of organic farming

1. Use of on-farm resources and avoiding/minimizing the use of off-farm resources.
2. Integrating farm resources for reducing the use of external inputs
3. Soil health management using farm wastes (crop, weed and animal wastes)
4. Generating feed resources in the farm: by growing forage crops and fodder (shrub and trees)
5. Green biomass generation for soil fertility management by growing high foliage producing plants and trees.
6. Composting of crop residues and livestock manures: with the help of earth worms converting them into value added vermi compost.
7. Preparing plant growth promoting and disease tolerant/ controlling substances (like Panchagavya, Coconut milk slurry, Amudhakaraisal *etc.*,)
8. Use of eco friendly pest management/crop protection measures (like use of predators and parasites, herbal extracts, crop diversification etc.,)

Objectives of organic farming

1. To produce food of high nutritional quality in sufficient quantity
2. To interact in a constructive and life enhancing way with all natural systems and cycles
3. To encourage and biological cycles with in the farming system, involving micro-organisms, soil flora and fauna, plants and animals and careful mechanical intervention
4. To maintain and increase long-term fertility of soils
5. To promote the healthy use and proper care of water, water resources and all life therein
6. To help in the conservation of soil and water
7. To use, as far as is possible, renewable resources in locally organized agricultural systems
8. To work, as far as possible, within a closed system with regard to organic matter and nutrient elements
9. To work, as far as possible, with materials and substances which can be reused or recycled, either on the farm or elsewhere
10. To give all livestock conditions of life which allow them to perform the basic aspects of their innate behaviour
11. To maintain all forms of pollution that may result from agricultural practices
12. To maintain the genetic diversity of the production system and its surroundings including the protection of wild life habitats
13. To allow everyone involved in organic production and processing a quality of life confirming to the UN Human Rights Charter, to cover their basic needs and obtain an adequate return and satisfaction from their work, including a safe working environment
14. To consider the wider social and ecological impact of thefarming system
15. To produce non-food products from renewable resources, which are fully degradable
16. Weed, disease and pest control relying primarily on crop rotation, natural predators, diversity, organic manuring, resistant

varieties and limited (preferably minimal) thermal, biological and chemical intervention

17. To create harmonious balance between crop production and animal husbandry
18. To encourage organic agriculture associations to function along democratic lines and the principle of division of powers
19. To progress towards an entire production, processing and distribution chain which is both socially just and ecologically responsible

Key Principles of Organic Agriculture

The International Federation for Organic Agriculture Movement's (IFOAM) definition of Organic agriculture is based on:

- The principle of health
- The principle of ecology
- The principle of fairness
- The principle of care

Each principle is articulated through a statement followed by an explanation. The principles are to be used as a whole. They are composed as ethical principles to inspire action.

Principle of health

Organic Agriculture should sustain and enhance the health of soil, plant, animal, human and planet as one and indivisible. This principle points out that the health of individuals and communities cannot be separated from the health of ecosystems - healthy soils produce healthy crops that foster the health of animals and people.Health is the wholeness and integrity of living systems. It is not simply the absence of illness, but the maintenance of physical, mental, social and ecological well-being.

Immunity, resilience and regeneration are key characteristics of health. The role of organic agriculture, whether in farming, processing, distribution, or consumption, is to sustain and enhance the health of ecosystems and organisms from the smallest in the soil to human beings. In particular, organic agriculture is intended to produce high quality, nutritious food that contributes to preventive health care and well-being. In view of this it should avoid the use of fertilizers, pesticides, animal drugs and food additives that may have adverse health effects.

Principle of ecology

Organic Agriculture should be based on living ecological systems and cycles, work with them, emulate them and help sustain them. This principle roots organic agriculture within living ecological systems. It states that production is to be based on ecological processes, and recycling. Nourishment and well-being are achieved through the ecology of the specific production environment. For example, in the case of crops this is the living soil; for animals it is the farm ecosystem; for fish and marine organisms, the aquatic environment. Organic farming, pastoral and wild harvestsystems should fit the cycles and ecological balances in nature. These cycles are universal but their operation is site-specific. Organic management must be adapted to local conditions, ecology, culture and scale. Inputs should be reduced by reuse, recycling and efficient management of materials and energy in order to maintain and improve environmental quality and conserve resources. Organic agriculture should attain ecological balance through the design of farming systems, establishment of habitats and maintenance of genetic and agricultural diversity. Those who produce, process, trade, or consume organic products should protect and benefit the common environment including landscapes, climate, habitats, biodiversity, air and water.

Principle of fairness

Organic Agriculture should build on relationships that ensure fairness with regard to the common environment and life opportunities. Fairness is characterized by equity,respect, justice and stewardship of the shared world, both among people and in their relations to other living beings. This principle emphasizes that those involved inorganic agriculture should conduct human relationships in a manner that ensures fairness at all levels and to all parties - farmers, workers, processors, distributors, traders and consumers. Organic agriculture should provide everyone involved with a good quality of life, and contribute to food sovereignty and reduction of poverty. It aims to produce a sufficient supply of good quality food and other products. This principle insists that animals should be provided with the conditions and opportunities of life that accord with their physiology, natural behavior and wellbeing. Natural and environmental resources that are used for production and consumption should be managed in a way that is socially and ecologically just and should be held in trust for future generations. Fairness requires systems of production, distribution and trade that are open and equitable and account for real environmental and social costs.

Principle of care

Organic Agriculture should be managed in a precautionary and responsible manner to protect the health and well-being of current and future generations and the environment. Organic agriculture is a living and dynamic system that responds tointernal and external demands and conditions. Practitioners of organic agriculturecan enhance efficiency and increase productivity, but this should not be at the risk ofjeopardizing health and well-being. Consequently, new technologies need to be assessed and existing methods reviewed. Given the incomplete understanding of ecosystems and agriculture, care must be taken. This principle states that precaution and responsibility are the key concerns in management, development and technology choices in organic agriculture. Science is necessary to ensure that organic agriculture is healthy, safe and ecologically sound. However, scientific knowledge alone is not sufficient. Practical experience, accumulated wisdom and traditional and indigenous knowledge offer valid solutions, tested by time. Organic agriculture should prevent significant risks by adopting appropriate technologies and rejecting unpredictable ones, such as genetic engineering. Decisions should reflect the values and needs of all who might be affected, through transparent and participatory processes.

Benefits of organic agriculture

Organic agriculture generates significant environmental and developmental benefits. It can contribute to meaningful socioeconomic and ecologically sustainable development, especially in poorer countries. This is due on the one hand to the application of organic principles, which means efficient management of local resources (e.g. local seed varieties, manure) and, therefore, cost-effectiveness. At the same time, the market for organic products – at the local and international level – has tremendous growth prospects and offers excellent opportunities to improve livelihoods for farmers all over the world.

1. Provides multifunctional benefits

In the past, the unsustainable production of food, feed, fibre and fuel strongly degraded global ecosystems and the services those systems provided for human survival. An area of 10 million hectares disappears by wind and water erosion every year and is therefore lost for food production, due to unsustainable farming techniques.

No other form of agriculture and food production can claim to offer so many benefits to consumers and to provide such a bounty of public goods as organic farming and food systems. Such ecosystem services include, for example:

- Provision of pure water,
- Recycling of organic matter and nutrients,
- Regulation of climate and weather events by fertile soils,
- Regulation of crop pests and diseases through biodiversity and natural enemies
- Pollination of crops by wild animals.

2. Biodiversity

The most notable environmental advantages may be summarized as follows: Organic farms grow several crops, including, trees, in carefully planned rotations or even as mixed cropping systems. Ideally, also husbandry is an integrated part of the farm system. The diversity not only allows optimum use of resources, but also serves as a form of economic security as it decreases the risk of vitality lost through pests, diseases, unfavorable weather or market conditions for certain crops.

It is not only the biodiversity of the produced crops and animals organic farms are aiming for, but also for the biodiversity of wild-life flora and fauna. A good proportion of the wild-fauna often consists of beneficials to control pests in the crops, thus are very useful helpers to assure and stabilize yield security. Providing and preserving a vital habitat for wild flora and fauna species—increasingly endangered and becoming extinct—is an extremely important and valuable service that sustainable agriculture provides that benefits society as a whole. It must be said, however, that handling and managing a wide range of biodiversity within crops and also with wild flora and fauna demands farmers with profound knowledge, highly professional skills and long-term experience.

3. Recycles plant nutrients

Organic nutrient management is based on biodegradable materials (i.e. plant and animal residues) that can be decomposed. Farms strive to create closed nutrient cycles whereby nutrients exported from the farm with the sold produce need to be replaced in some way with the help of composting, mulching, green manuring, crop rotation and cultivation of nitrogen fixing plants. Farm animals also play an important role in the nutrient cycle: their waste is of high value and its use allows recycling nutrients provided with the fodder. If carefully managed, losses of nutrients due to leaching, soil erosion and through gases can be reduced to a minimum. These measures mentioned minimize the need and the dependency on external nutrient inputs and help to save significant production expenses.

4. Employ, natural and biological control of pests and diseases

Organic farmers try to keep pests and diseases at a level which does not cause economical damage through a concept of several preventive measures. The main focus supports the vigour and robustness, or self-defence potential by cropping through careful management. Resistant or tolerant cultivars are used wherever they exist and fulfil market requirements; beneficial insects are promoted by offering them a favourable habitat and food sources. If pests reach critical population levels to decrease yields significantly, natural preparations and biocontrol agents and methods are applied as control measures.

5. Integrates animal husbandry into the production system

Organic farmers, when possible, integrate farm animals into their production system to support the recycling of nutrients, obtain animal products for household nutrition and sale to optimize the family income. The animals must be provided with the conditions and opportunities of life in accordance with their physiology and natural behaviour (e.g. stable construction and herd management) The health of farm animals is ensured by primarily selecting strong and locally adapted breeds, followed by providing a balanced nutrition, clean and safe housing, continuous monitoring and using natural means for disease and parasite control.

6. Improves and conserve soil conditions

Organic farmers give central importance to the improvement and conservation of soil conditions. They protect the topsoil as well as organic matter in the soil from loss through soil erosion control, mulching, cover cropping, green manuring, application of compost, adequate mechanisation and management methods to avoid soil compaction and degradation and minimum tillage practices. All these measures improve and stabilize the soils physical structure, enhance its ability to absorb and store water and plant nutrients and stimulate the activity of soil organisms, roots and finally plant performance.

7. Lower negative environmental impacts

The high dependence of traditional farming on chemical fertilizers, herbicides and pesticides has caused considerable environmental damage. Due to the ban of chemical fertilizers on organic farms, 35 to 65 percent less nitrogen leaches from arable fields into soil zones where it could degrade ground and drinking water quality. Other nutrient elements like potassium and phosphorous are not found in excessive quantities in organic soils, which increases their efficient use.

Since synthetic herbicides and pesticides are not applied in organic farms, they cannot be found in their soils, surface and ground water.

8. Better Water Quality

As agriculture is dependent upon water availability, agricultural practices in turn impact water quality. Research has shown that organic farming contributes to cleaner water by using biological fertilizers that release nutrients slowly, build soil organic matter, increase soil water-holding capacity and reduce leaching of nitrates into groundwater.Ground and surface waters are too often contaminated by pesticides, fertilizers and animal wastes that are not absorbed by plants or soil.

9. Stable soils – less prone to erosion

Fertile soils with stable physical properties have become the top priority of sustainable agriculture. Essential conditions for fertile soils are vast populations of bacteria, fungi, insects and earthworms, which build up stable soil aggregates. There is abundant evidence that organic farms and organic soil management lead to good soil fertility. Compared to conventionally managed soils, organically managed ones show higher organic matter contents, higher biomass, higher enzyme activities of microorganisms, better aggregate stability, improved water infiltration and retention capacities. Finally less water and wind erosion. The fact that organic farmers use a plough periodically in order to bury weed roots and seeds, does not render their soils more prone to erosion.

10. Carbon sequestration

Organic agriculture contributes to mitigating the greenhouse effect and global warming through its ability to sequester carbon in the soil. Many management practices used by organic agriculture (e.g. minimum tillage, returning crop residues to the soil, the use of cover crops and rotations and the greater integration of nitrogen-fixing legumes), increase the return of carbon to the soil, raising productivity and favoring carbon storage. A number of studies revealed that soil organic carbon contents under organic farming are considerably higher. The more organic carbon is retained in the soil, the more the mitigation potential of agriculture against climate change is higher.

Organic farmers use different techniques for building up soil fertility. The most effective ones are fertilization by animal manure, by composted harvest residues and by leguminous plants as (soil) cover and (nitrogen) catch crops. Introducing grass and leguminous leys as feedstuff for ruminants into the rotations and diversifying the crop sequences, as well as reducing ploughing depth and frequency, also augment soil fertility. All these techniques also increase carbon sequestration rates on organic fields. A further increase of carbon capture in organically managed fields can be measured by reducing the frequency of soil tillage.

Sustainability of organic farming

The goal of organic agriculture is to contribute to the enhancement of sustainability. In the context of agriculture, sustainability refers to the successful management of agricultural resources to satisfy human needs while at the same time maintaining or enhancing the quality of the environment and conserving natural resources for future generations. Sustainability in organic farming must therefore be seen in a holistic sense, which includes ecological, economic and social aspects

The organic agriculture techniques are known to be ecologically sustainable by:

- Improving soil structure and fertility through the use of crop rotations, organic manure, mulches and the use of fodder legumes for adding nitrogen to the soil fertility cycle.
- Prevention of soil erosion and compaction by protecting the soil planting mixed and relay crops.
- Promotion of biological diversity through the use of natural pest controls (e.g. biological control, plants with pest control properties) rather than synthetic pesticides which, when misused, are known to kill beneficial organisms (e.g. natural parasites of pests, bees, earthworms), cause pest resistance, and often pollute water and land.
- Performing crop rotations, which encourage a diversity of food crops, fodder and under-utilized plants; this, in addition to

improving overall farm production and fertility, may assist the on-farm conservation of plant genetic resources.

- Recycling the nutrients by using crop residues (straws, stovers and other non-edible parts) either directly as compost and mulch or through livestock as farmyard manure.
- Using renewable energies, by integration of livestock, tree crops and on farm forestry into the system. This adds income through organic meat, eggs and dairy products, as well as draught animal power. Tree crops and on-farm forestry integrated into the system provide food, income, fuel and wood.

Social sustainability

Sustainability is also about equity among and between generations. Organic agriculture contributes to the social well-being by reducing the losses of arable soil, water contamination, biodiversity erosion, GHG emissions, food losses, and pesticide poisoning.

Organic agriculture is based on traditional knowledge and culture. Its farming methods evolve to match localenvironments, responding to unique biophysical and socio economics constraints and opportunities. By using local resources, local knowledge, connecting farmers, consumers and their markets, the economic conditions and the development of rural can be improved.

Organic agriculture stresses diversification and adaptive management to increase farm productivity, decrease vulnerability to weather vagaries and consequently improves food security, either with the food the farmers produce or the income from the products they sell.

Economic sustainability

Organic farming appears to generate 30% more employment in rural areas and labor achieves higher returns per unit of labor input. By using local resources better, organic agriculture facilitates smallholders' access to markets and thus income generation; and relocalizes food production in market-marginalized areas.

Generally, organic yields are 20% less as compared to high-input systems in developed countries but could be up to 180% higher as compared to low-input systems in arid/semi-arid areas. In humid areas, rice paddy yields are equal, while the productivity of the main crop is reduced for perennials, though agroforestry provides additional goods.

Operating costs (seeds, rent, repairs and labor) in organic agriculture are significantly lower than conventional production, ranging from 50-60% for cereals and legumes, to 20-25% for dairy cows and 10-20% for horticulture products. This is due to lower input costs on synthetic inputs, lower irrigation costs, andlabor cash costs that include both family labor and hired workers. Total costs are, however, only slightly lower than conventional, as fixed costs (such as land, buildings and machinery) increase due to new investments during conversion (e.g., new orchards, animal houses) and certification.

Organic versus conventional farming

Organic Farming Vs Conventional Farming

Conventional farming	Organic farming
It is based on *economical* orientation, heavy mechanization, specialization and misappropriates development of enterprises with unstable market oriented programme	It is based on *ecological* orientation, efficient inputuse efficiency, diversification and balanced enterprise combination with stability
Supplementing nutrients through fertilizers, weed control by herbicides, plant protection measures by chemicals and rarely combination with livestock	Cycle of nutrients within the farm, weed control by crop rotation and cultural practices, plant protection by non-polluting substances and better combination of livestock
Based on philosophy of to feed thecrop/plants	"Feed the soil not to the plant" is the watch word and slogan of organic farming
Production is not integrated into environment but extract more through technical manipulation, excessive fertilization and nocorrection of nutrient imbalances	Production is integrated into environment, balanced conditions for plants and animals and deficiencies need to be corrected
Low input : output ratio with considerable pollution	High input : output ratio with no pollution
Economic motivation of natural resources without considering principles of natural up gradation	Maximum consideration of all natural resources through adopting holistic approaches

In contrast, several agro-ecologically based researchers stress more the fluid transition between conventional, integrated and organic farming, as an outcome of different assessments of economic, ecological and social goals. Consequently, technique strategies such as integrated pest management of balanced nutrient supply might improve conventional agriculture to such as an extent that it may appear unnecessary to strictly ban pesticides and mineral fertilizers as required by organic standards.

However, there is scientific that organic agriculture differs from conventional agriculture not only gradually but fundamentally. Implementing organic methods consequently seems to provide a new quality in how the agro-ecosystem works. This functioning cannot be explained by summing up single ecological measures. Organic farming seems to improve soil fertility in a way and to an extent which cannot be achieved by conventional farming even if the later consistently respects some ecologically principles.

Organic agriculture is one of several to sustainable agriculture and many of the techniques used (e.g. inter-cropping, rotation of crops, double digging, mulching, integration of crops and livestock) are practiced under various agricultural systems. What makes organic agriculture unique, as regulated under various laws and certification programmes, is that:

1) Almost all synthetic inputs are prohibited and
2) Soil building crop rotations are mandated.

The basic rules of organic production are that natural inputs are approved and synthetic inputs are prohibited, but there are exceptions in both cases.

Certain natural inputs determined by the various certification programmes to be harmful to human health or the environment are prohibited (e.g. arsenic). As well, certain synthetic inputs determined to be essential and consistent with organic farming philosophy, are allowed (e.g. insect pheromones). Lists of specific approved synthetic inputs and prohibited natural inputs are maintained by all the certification programmes. Many certification programmes require additional environmental protection measures in adoption to these two requirements. While many farmers in the developing world do not use synthetic inputs, this alone is not sufficient to classify their operations as organic.

Components of organic farming

Essential components of organic farming are keeping the soil alive through effective management natural resources. They are as follows:

- **Enrichment of soil:** Abandon use of chemicals, use crop residue as mulch, use organic and biological fertilizers, adopt crop rotation and multiple cropping, avoid excessive tilling and keep soil covered with green cover or biological mulch.
- **Management of temperature**: Keep soil covered, plant trees and bushes on bund
- **Conservation of soil and rainwater:** Dig percolation tanks, maintain contour bunds in sloppy land & adopt contour row cultivation, dig farm ponds, maintain low height plantation on bunds.
- **Harvesting of sun energy:** Maintain green stand throughout the year through combination of different crops and plantation schedules.
- **Self-reliance in inputs:** Develop your own seed, on-farm production of compost, vermicompost, vermiwash, liquid manures and botanical extracts. Preparing plant growth promoting and disease tolerant/ controlling substances (like Panchagavya, Coconut milk slurry, Amudhakaraisal *etc.,*)
- **Maintenance of life forms**: Develop habitat for sustenance of life forms, never use pesticides and create enough diversity.
- **Integration of animals**: Animals are important components of organic management and not only provide animal products but also provide enough dung and urine for use in soil.
- **Use of renewable energy**: Use solar energy, bio-gas and other eco-friendly machines.

Organic conversion challenges

A. Climate-related challenges to conversion

Converting a farm to organic farming in an area with very little rainfall and high temperatures or strong winds will be more challenging than converting a farm located in an area with well distributed rainfall and favourable temperatures. At the same time, the improvements that follow implementation of organic practices will be more obvious under arid conditions than under ideal humid conditions. For example, compost application into topsoil or into planting holes will increase the soils water retention capacity and the crop's tolerance to water scarcity.

In very warm and dry climate, losses of water through transpiration from plants and evaporation from soils are high. These losses may be further encouraged by strong winds, enhancing soil erosion. The soils' organic matter content is generally low, as biomass production is low, implying that the availability of nutrients to the plants is highly reduced. Under such conditions, the key to increasing crop productivity lies in protecting the soil from strong sun and wind and increasing the supply of organic matter and water to the soil. Soil organic matter can either be increased through compost or through cultivation of green manure crops. In the case of compost production the challenge is to increase production of plant biomass, which is needed for compost production.

In warm and humid climate, high aboveground biomass production and rapid decomposition of soil organic matter imply that the nutrients are easily made available to the plants. But it also involves a high risk that the nutrients are easily washed out and lost. Under such conditions a balance between production and decomposition of organic matter is important to avoid depletion of soil. Combining different practices to protect the soil and feed it with organic matter proves to be the most effective approach to choose. These practices include creating a diverse and multi-layer cropping system ideally including trees, growing nitrogen-fixing cover crops in orchards and applying compost to enrich the soil with organic matter and in this way increase its capacity to retain water and nutrients.

B. Social and cultural challenges to conversion

In addition to agricultural and ecological aspects, also social and cultural aspects are relevant in conversion to organic farming. In most Indian cultures, farming is communal and highly regarded as a social activity, whereby decisions regarding what, how and where to grow is taken by either the whole family or the community. So changes in farming, such as the introduction of organic farming practices, needs to be discussed with the family and the community. Key aspects to consider include the ideas of family members about conversion to organic farming, their aims and expectations. A farming family or community needs to sit together to agree on what they wish to achieve through conversion to organic farming. Points to take into consideration include income, availability of food for own consumption, the amount of firewood produced on the farm and the workload of each family member.

C. Economic challenges to conversion

The decision to farm organically is in most cases a commitment for the future of farming: Commonly, when farmers and their families decide to convert, they aim to improve their incomes and livelihoods. In a first period of the conversion process, however, some investments may be required. Such investments may include, for example, purchasing of appropriate equipment for soil cultivation, for weed control or for compost production. It may also be recommended to buy animals or specific seeds in order to diversify production. Improvements may be necessary for housing animals, storing manures or storing farm products.

Furthermore, additional labour may be needed for constructing erosion control structures, for composting. Further, the decision to become an organic farmer also includes the decision to improve efforts in marketing. Conversion to organic agriculture also requires time and investment in building knowledge and in setting up a marketing infrastructure, for example building an on-farm store or finding new buyers. However, all the above requirements will vary with the size of the farm, the intensity of production and the market channels being targeted.

Which crops to grow during conversion?

Looking at the organic farm as being 'one organism', the focus does not lie on cultivating specific crops only. Rather, the focus is on choosing crops that can easily be integrated into the existing farming system and will contribute to its improvement. But the choice also depends on the farmer's knowledge on the right management of the crops, their contribution to a diverse family diet or their demand in the market. Besides growing crops for food, farmers may need to grow leguminous cover crops to provide high-protein feed for livestock and to be used as green manures to feed the soil. Planting trees for shade, as windbreak, for firewood, feed, mulching material or for other uses, can be recommended in most situations.

Criteria for crop selection during conversion

- In a first place organic farmers should grow enough food for the family. But they may also want to grow crops for the market to get money for other family needs. The farmers should also grow crops that contribute to improvement of soil fertility. Farmers who keep livestock need to grow pasture grass and legumes.
- Basically, farmers should select crops with low risk of failure. Cereals and legumes such as maize, sorghum, millet, beans and peas are especially suitable for conversion, since they cost little to produce, generally have moderate nutrient demands and are robust against pests and diseases. In addition, many of the traditional crops can be stored and sold in domestic markets. High-value short term crops, such as most vegetables, are more delicate to grow and highly susceptible to pest and disease attack. Therefore, they should not be grown on a larger scale, unless the farmer can endure some losses in harvest.
- The crops to grow for sale should include crops that can be sold at the farm gate, at the roadside market or can be transported directly to nearby markets in urban centres. Choosing the right crop to sell on the market may require some market information. Decision making for crops for local or export markets requires detailed information from traders or

exporters on the crops, requested varieties, quantities, qualities and season.

- High-value perennial crops such as fruit trees take at least 3 years until the first harvest from the date of planting. This makes them appropriate crops for the conversion period. For new plantations, species and varieties must be carefully selected to suit the organic market and production requirements. For conversion of an existing orchard, it might be necessary to replace old existing varieties, if they are very susceptible to diseases and the product quality does not match with the market requirements.
- The success of a crop will also depend on provision of favourable growing conditions. The better a crop variety matches local soil and climate conditions and is tolerant or resistant to common pests and diseases, the better it will grow.
- Planting of hedges and agroforestry trees can be valuable to help establish a diverse farming system.
- Growing leguminous green manures provides nutrients to the soil. Green manures do not provide immediate income, but in the long-term, they make the soil fertile and productive for the future.

Characteristics of an ideal organic farm

Organic agriculture aims at successfully managing natural resources to satisfy human needs while maintaining the quality of the environment and conserving resources. Organic agriculture thus aims at achieving economic, ecological and social goals at the same time.

a. The ecological goal

The ecological goal basically relates to maintenance of quantity and quality of natural resources. Farming should be done in an environmentally-friendly manner, whereby the soil, water, air, plants and animals are protected and enhanced. Organic farmers pay special attention to the fertility of the soil, the maintenance of a wide diversity of plants and animals, and to animal friendly husbandry.

Important environmental goals are:

- Prevention of loss and destruction of soil due to erosion and compaction.
- Increasing the humus content of the soil.
- Recycling farm-own organic materials and minimizing use of external inputs.
- Promotion of natural diversity of organisms - being a criterion of a balanced natural ecosystem.
- Prevention of pollution of soil, water and air.
- Ensuring husbandry that considers natural behaviour of farm animals.
- Use of renewable energy, wherever possible.

To achieve these goals organic farmers maintain wide crop rotations, practice intercropping and cover cropping, plant hedgerows and establish agro-forestry systems.

b. The social goal

Organic farming aims at improving the social benefits to the farmer, his/her family and the community in general.

Important social goals include:

- Creating good working conditions for all.
- Ensuring a safe nutrition of the family with healthy foods.
- Ensuring sufficient production for subsistence and income.
- Encouraging fair and conducive working conditions for hired workers.
- Encouraging learning and application of local knowledge.

From an organic perspective, at the household level fair participation in farm activities of all family members and proper sharing of the benefits from the farm activities is essential. On community level, knowledge and experiences should be shared and collaboration strengthened in order to obtain higher benefits.

c. The economic goal

In an economic sense organic farming aims at optimizing financial benefits to ensure short- and long-term survival and development of the farm. An organic farm should not only pay for production costs, but also meet the household needs of the farmer's family.

Important economic goals include:

- Satisfactory and reliable yield.
- Low expenditures on external inputs and investments.
- Diversified sources of income for high income safety.
- High value added on-farm products through improvement of quality and on-farm processing of products.
- High efficiency in production to ensure competitiveness.

Organic farmers try to achieve this goal by creating different sources of income from on- and off-farm activities. Usually different crop and animal enterprises are adopted simultaneously in a mixed production system. The target also includes being more self-sufficient in terms of seeds, manures, pesticides, food, feeds, and energy sources and thereby minimizing cash outlay to purchase off-farm items.

Prospects of organic farming

Deficiencies of at least five out of the critical soil nutrients are widespread in Asia due to imbalance in application of fertilizers and very limited use of organic manures. Organic techniques alone can help to regenerate the degraded soils and ensure sustainability in crop production.

Soil organic matter is the life source of dynamic soil. The decline in soil organic matter in the recent times in Indian soils often associated with crop yield loss of about 30 per cent and organic manures alone can sustain the productivity of the soil. There is an increase in crop yield by 12 % for every 1% increase in organic matter, as evident from long term manurial experiment conducted at India.

As organic farming is attracting worldwide attention, and there is a potential for export of organic agricultural produce, this opportunity has to be tapped with adequate safeguards so that the interest of small and marginal farmers can be well protected.

Organic farming may be practiced in crops, commodities and regions where the country has comparative advantage. To begin with, the practice of organic farming should be for low volume high value crops like spices, medicinal plants, fruits and vegetables.

Besides the identification of regions suitable for the adoption of organic farming, the crops and their products should also be identified which are amenable for production through organic ways and have the potential to fetch a premium price in the international organic market.

Organic farming should not be confined to the age old practice of using cattle dung, and other inputs of organic/biological origin, but an emphasis needs to be laid on the soil and crop management practices that enhance the population and efficiency of belowground soil biodiversity to improve nutrient availability. Performance of cultural techniques for weed control and that of bio-pesticides for pest management need to be evaluated under field conditions, preferably under cultivators.

Indian agricultural activity results in abundant crop residues. The residue turnover is 273.63 mt and the nutrient potential is 5.67 mt of NPK. Proper residue recycling can serve as effective substitute for inorganic fertilizers. Large potential of organic resources remain untapped in India. Nearly 750 mt cow dung and 250 mt of buffalo manures are available.

Crop rotation including pulses/green manures, which fixes atmospheric nitrogen and leave root nodules in the soil and help in improving residual nitrogen content thereby economizing nitrogen use.

Use of forest leaf litters, in places of availability (For example: Andaman and Nicobar Islands) greatly improvises the soil organic matter and in turn soil organic carbon status.

Integration of traditional knowledge with scientific non chemical inputs and methods brings sustainability in farming.

Although, commercial organic agriculture with its rigorous quality assurance system is a new market controlled, consumer-centric agriculture system world over, but it has grown almost 25-30% per year during last 10 years. In spite of recession fears the growth of organic is going unaffected. The movement started with developed world is gradually picking up in developing countries. But demand is still concentrated in developed and most affluent countries. Local demand for organic food is growing. India is poised for faster growth with growing domestic market. Success of organic movement in India depends upon the growth of its own domestic markets.

India has traditionally been a country of organic agriculture, but the growth of modern scientific, input intensive agriculture has pushed it to wall. But with the increasing awareness about the safety and quality of foods, long term sustainability of the system and accumulating evidences of being equally productive, the organic farming has emerged as an alternative system of farming which not only address the quality and sustainability concerns, but also ensures a debt free, profitable livelihood option.

Chapter 2

Scenario of Organic Farming

Agriculture in India is not of recent origin, but has a long history dating back to Neolithic age (7500–6500 B.C). It changed the life style of early man from nomadic hunter of wild berries and roots to cultivator of land. Agriculture is benefited from the wisdom and teachings of great saints. The wisdom gained and practices adopted have been passed down through generations. The traditional farmers have developed the nature friendly farming systems and practices such as mixed farming, mixed cropping, crop rotation etc. The great epics of ancient India convey the depth of knowledge possessed by the older generations of the farmers of India. The modern society has lost sight of the importance of the traditional knowledge, which had been subjected to a process of refinement through generations of experience. The ecological considerations shown by the traditional farmers in their farming activities are now a days is reflected in the resurgence of organic agriculture.

Historical Perspective of Organic Farming

The available ancient literature includes the four Vedas, nine Brahnanas, Aranyakas, Sutra literature, Susruta Samhita, Charaka Samhita, Upanishads, the epics Ramayana and Mahabharata, eighteen Puranas, Buddhist and Jain literature, and texts such as Krishi-Parashara, Kautilya's, Artha-sastra, Panini's Ashtadhyahi, Sangam literature of Tamils, Manusmirit, Varahamihira's Brahat Samhita, Amarkosha, Kashyapiya-Krishisukti and Surapala's Vriskshayurveda. This literature was most likely to have been composed between 6000 B.C. and 1000 A.D. The information related to the biodiversity and agriculture (including animal husbandry) is available in these texts.

Rig-Veda is the most ancient literary work of India. It believed that Gods were the foremost among agriculturists. According to 'Amarakosha', Aryans were agriculturists. Manu and Kautilya prescribed agriculture, cattle rearing and commerce as essential subjects, which the king must learn. According to Patanjali the economy of the country depended on agriculture and cattle-breeding. Plenty of information is available in 'Puranas', which reveals that ancient Indians had intimate knowledge on all agricultural operations. Some of the well known ancient classics of India are namely, Kautilya's'Arthashastra'; Panini's 'Astadhyayi'; Patanjali's 'Mahabhasya'; Varahamihira's 'Brahat Samhita'; Amarsimha's 'Amarkosha' and Encyclopedic works of Manasollasa. These classics testify the knowledge and wisdom of the people of ancient period. Technical book dealing exclusively with agriculture was Sage Parashara's 'Krishiparashara' in 1000 A.D. Other important texts are Agni Purana and Krishi Sukti attributed to Kashyap (500 A.D.).

The Mahabharata mentions of Kamadhenu, the celestial cow and its role on human life and soil fertility.

Ancient Tamil and Kannada works contain lot of useful information on agriculture in ancient India. Agriculture in India made tremendous progress in the rearing of sheep and goats, cows and buffaloes, trees and shrubs, spices and condiments, food and non-food crops, fruits and vegetables and developed nature friendly farming practices. These practices had social and religious undertones and became the way of life for the people. Domestic rites and festivals often synchronized with the four main agricultural operations of ploughing, sowing, reaping and harvesting. In the *Rig-Veda,* there is reference to hundreds and thousands of cows; to horses yoked to chariots; to race courses where chariot races were held; to camels yoked to the chariots; to sheep and goats offered as sacrificial victims, and to the use of wool for clothing. The famous Cow Sukta (Rv. 6.28) indicates that the cow had already become the very basis of rural economy. In another Sukta, she is defined as the mother of the Vasus, the Rudras and the Adityas, as also the pivot of immortality. The Vedic Aryans appear to have large forests at their disposal for securing timber, and plants and herbs for medicinal purposes appear to have been reared by the physicians of the age, as appears in the *Atharva Veda.* The farmers' vocation was held in high regard, though agriculture solely depended upon the favours of Parjanya, the god of rain. His thunders are described as food-bringing.

Agriculture during the sangam age

Agriculture was the Principal occupation of Tamils. The Agriculturalists were called ' Ulavar' and their women the 'Ulattiyar'. The classes of people owning land and the class of people actually tilling the land were 'Vellalas' the farmer known as the superior 'Vellalas' and the latter known on inferior 'Vellalas. Ulavar was also known as Valnar. Ulutunbar or Yerin. Purananuru calls Ulavar as Kalamar. The term Ulavar itself indicates the use of the plough and the term Vellalar denotes the propertiership of the soil. The cowherd community counted the cattle as wealth while among agriculturalists the number of plough was the standard of measurement of wealth. Thiruvalluvar had highlighted the importance of Agriculture in PART-104, Thirukural. Agriculture is considered as an esteemed profession. Valluvar had described the desirable feature of a territory or country. A country should have good agriculturalists and learned and wealthy men. It must be free from hunger, disease, and enmity. A country should not be under the influence of famine. In whatever occupation others might be engaged they might engaged, they must all depend finally on the farmer, even the ascetics will become helpless if presents do not till the lands. Agriculture is not as dignified as other professions; on the other hand, the agriculturalists are positively the support of whole world. Agriculturalists alone lead a truly useful life, the rest being only parasites and sycophants. According to Thiruvalluvar an agriculturalist must: (*i*) plough the land; (*ii*) manure it; (*iii*) transplant the seedlings; (*iv*) ensure an unfailing supply of channel water and (*v*) protect the cultivated farm from the stray cattle. He warns the farmer against lethargy, he lids him be active and never despond. The farmer is to guard against absentee-landlordism.

Agricultural Implements

Buffaloes were used for ploughing with a wooden plough. Deep ploughing was considered superior to shallow ploughing. A labour saving tool called parambu was used for levelling paddy fields. Tools such as amiry, keilar, and yettam were used to lift water from wells, tanks, and rivers. Tools called thattai and kavan were used for scaring birds in millet fields. Traps were used to catch wild boars in millet fields.

Land Preparation

Thiruvalluvar gives a few ideas about agricultural operations. If an agriculturalist would allow the ploughed land to dry up so that one todi (one palam) of dust dries down to one kashi (1/4 palam) *i.e.*, if it is reduced to one fourth (1/4) of the original quantity, there will be no need to put into the land even one handful of eru, *i.e.,* manure. Ploughing was carried out many times instead of single time. Agriculture can be practiced easily when the cultivator has his own ploughs. Ploughing one time was referred as Orusal ulavu; twice as Irusal ulavu and many times as Chensal ulavu. Plough the land deeper than wider. Cattle were used for ploughing. Cyperus weeds and crab cavities were destroyed during land preparation and levelled in wetlands. Crops had been raised in beds and channels.

Crops and Varieties

Ancient Tamils cultivated paddy, black gram, horse gram, varagu, tenai, sesame, sugarcane, banana, coconut, palmgrab, bamboo grasses, jack fruit, tamarind and mango. Varagu was cultivated in Mullai lands. Tenai and field bean (Mochai) were cultivated as mixed crop in Kurinchi lands. Cotton and Tenai were cultivated as mixed crops. Rice varieties such as Chennel, Vennel, Salinel, Mudandainel, Ivananel, Kulanel, Thoppinel, Pudunel, Varnel, Aviananel, Torainel were cultivated. Mungil el or Mungil arisi obtained from bamboo. It was taken as food at the time of king Pari. Red gram, Black gram was cultivated in Marutham lands. Sugarcane was cultivated with check basin Method at the foot hills. Sugarcane var. Kalik karumbu was cultivated in Thagadur region during the king Adiyaman period. Banana was cultivated. Rice, sugarcane, coconut, plantains, areca palm, turmeric, mango, palmyra, sembu (*Colocasia antiquolam*) and ginger were grown in Cauvery river valley. A 'Veli' of land produced a round thousands kalams of paddy. Farmer enjoys on seeing the first flushes and hearing the sound of Cauvery flow and of the eddying water scouring the bunds.

Seeds and Sowing

Seed was selected from those earheads that first matured. The selected seed was stored for sowing only and never used as food grain. It was believed that such a diversion would destroy the family. Sowing tenai seeds without ploughing was also practiced. Cyperus weeds were removed through feeding with pigs and then in such lands seeds were sown without ploughing. Seeds were sown with adequate spacing. Seed germination happens with adequate moisture.

Cropping Systems

Crop rotation was practiced by raising black gram (urd) after rice. They also practiced mixed cropping; *e.g.*, foxtail millet with lablab or cotton. Ginger and turmeric were grown as intercrops in coconut and jack fruit plantations. Rice fallow cultivation with other crops such as pulses had been reported in 'Ingurunuru'. Cultivation of sugarcane was reported in 'Pathittrupattu'. Mixed cropping of cotton and tenai were also practiced. Pepper was grown as mixed crop in mango plantation.

Weed Management

Weeds were removed from the fields. Tools were used for weed control. Weeds hamper the growth of crops.

Soil Fertility

Thiruvalluvar stated that fertile land alone is entitled to be called territory which yield wealth unsought for. The fertility of the land especially in Chola Kingdom finds proud mention in contemporary literature. Organic manures were applied in ploughed lands. Avur Mulankilar in a short poem addressed to Killivalavan says, "The land is so fertile that a tiny piece there of, where a she-elephant might rest, can produce enough food to nourish seven bull elephants.

The fertility of the land even in hilly areas like the Palakunrakottam (land between Tirupathi and Tiruvannamalai) was such that the sesame crop was so healthy and full grown that a handful could contain no more than seven grains of sesame. Even without ploughing, merely sowing deep in turned sod made mustard grown in great quantities. In a fairly fertile farm, a veli of land produced a full thousand kalams of paddy. The silt carried by the flood water was a major source of fertilization, and the greater the volume of water, the greater the valuable silt deposit. Some of the more favourably situated fields were known as "Erikkattu" meaning tank reservoir. This was an ingenious system of "field insurance" against the risk of floods.

Irrigation Management

Water quality depends on land type. Moisture stressed crops grow well on receipt of rains; construction of ponds for others use is essential.

(i) **Art of well divining -** The Cankam art works speak of the art of well-divining practice of the Tamils to wells on the highways for the weary travellers. Although 'Kuval, Acumpu, Kupam, Kuli, Puval, Keni, Turavu are used synonymously to denote well, cankam classic speak of kuval only. According to two manuscripts, rocky lands were classed as 'Kurinchi'; the land with coarsesand, 'Neytal'; the land abounding in scattered minor rocks, 'Mullai'; muddy land, 'Marutam' and the unused tract 'Palai', of which the Neytal tract was supposed to have moisture. The depths of the water source in different lands differed from the surface land. In Kurinchi springs will be found at a depth of 33 cans, in Palai 30 cans, in Mullai 36 cans, in Neytal 35 cans and in Marutam 22 cans. The soil fit for the growth of banyan tree, tamarind, mango and so on might have water springs at different depths. The places where white rats, scorpion, the double-tongued lizard, toad and so on inhabit might have water sources. Another manuscript talks about the brownishness of Mullai water, whiteness of Palai water, Kurinchi's blackish water, Marutam's potable water and saltish water of Neytal. A well, which had disappeared due to human or natural calamities, can be traced on the basis of certain varieties of grass getting withered in winter and flourishing in summer season. In such places, there would be a swarm of flies and ants; also anthills appearing in places where certain grassy plants grow and such trees as 'Vanci' and 'Nocci' flourishing during the hot summer season, would be sure indications of the existence of wells-now disappeared in such places.

(ii) **Major irrigation system of ancient Tamil Nadu** - The Pallavas, whose capital was in Tondaimandalam, constructed several irrigation tanks, and practically all of them are functioning to this day. The Cholas, besides constructing tanks, tamed the Cauvery river, an achievement of which any monarch and his people may be proud. The Pandyan Country was divided between fringe irrigation alongside of the rivers and the utilization of tanks. The two major rivers of Tamil Nadu are the Cauvery and the Tamiraparani. The Vaigai has at no time been a source of great importance. The Cauvery river rises in the western ghats near Coorg and after a course of nearly 500 miles, enters the Bay of Bengal, draining an area of about 31,000 square miles in route. In the Cholamandala Satakam there is mention of the Kallanai or the Grand Anicut being constructed. Karikalan is said to have employed several thousands of Ceylonese for this purpose. According to the "Mahavamsa", one hears of an aged woman complaining to Gajabahu that amongst the twelve thousand persons taken away by Karikalan for making the embankment of the Cauvery was her only son. According the Pattinathupalai, Karikalan was known as "Kaaverinaadan" due to his taming the violent river. His raising of the flood banks of the Cauvery was mentioned in the Malepadu plates of Punayakumara, a Telegu Choda king of the seventh or eighth century.

Plant Protection

Fencing had been laid out around the fields to protect from the animals. Fencing was done with bamboo

Harvesting and Threshing

Harvesting was carried out in night time with beating drums to protect from the wild animals. Rice crop was harvested using a tool 'Kuyam'. Rice was threshed using cattle and elephants. Garden land bean (Avarai) was cultivated in Tenai stubbles; sowing of tenai and cotton in harvested tenai lands. A tool called 'senyam' was used for harvesting rice. Threshing of rice was done by hand with the help of a buffalo (and in large holdings by elephants). Hand winnowing was done to remove chaff. One sixth of the produce was paid as tax to the king. Farm labourers were paid in kind. The land was immediately ploughed after harvest or water was allowed to stagnate to facilitate rooting of stubble. Operations requiring hard work such as ploughing were done by men while women attended to light work such as transplanting, weeding, bird scaring, harvesting and winnowing. In Kandapuranam, it is mentioned that Valli, the daughter of a king, was sent for bird scaring in millet fields where Lord Muruga (son of Lord Shiva) courted her and married.

Developmental Era of Organic Farming

The development of the organic farming era worldwide had gone through mainly three stages, Emergence, Development, and Growth in chronological sequence.

Era of Emergence (1924–1970) - Pre-World War II

The beginning of organic farming could trace back to 1924 in Germany with Rudolf Steiner's course on *Social Scientific Basis of Agricultural Development*, in which his theory considered the human being as part and parcel of a cosmic equilibrium that he/she must understand in order to live in harmony with the environment.

The first 40 years of the 20th century saw simultaneous advances in biochemistry and engineering that rapidly and profoundly changed farming. The introduction of the gasoline powered internal combustion engine ushered in the era of the tractor and made possible hundreds of mechanized farm implements. Research in plant breeding led to the commercialization of hybrid seed. And a new manufacturing process made nitrogen fertilizer - first synthesized in the mid-19th century - affordably abundant. These factors changed the labour equation.

Consciously organic agriculture (as opposed to the agriculture of indigenous cultures, which always employs only organic means) began more or less simultaneously in Central Europe and India. The

British botanist Sir Albert Howard is often referred to as the father of modern organic agriculture. From 1905 to 1924, he worked as an agricultural adviser in Pusa, Bengal, where he documented traditional Indian farming practices and came to regard them as superior to his conventional agriculture science. His research and further development of these methods is recorded in his writings, notably, his 1940 book, *An Agricultural Testament*, where he developed the famed Indore composting process, which put the ancient art of composting on a firm scientific basis and explained the relationship betwenn the health of the soil, the health of plants and the health of animals.

In 1909, American agronomist F.H. King toured China, Korea, and Japan, studying traditional fertilization, tillage, and general farming practices. He published his findings in *Farmers of Forty Centuries* (1911, Courier Dover Publication). King foresaw a "world movement for the introduction of new and improved methods" of agriculture and in later years his book became an important organic reference.

In Germany, Rudolf Steiner's biodynamic agriculture, was probably the first comprehensive organic farming system. This began with a lecture series Steiner presented at a farm in Koberwitz (now in Poland) in 1924. Steiner emphasized the farmer's role in guiding and balancing the interaction of the animals, plants and soil. Healthy animals depended upon healthy plants (for their food), healthy plants upon healthy soil, healthy soil upon healthy animals (for the manure).The lectures were compiled as *The Agriculture Course* in 1928.Because of Steiner the first certificate on natural farming namely Demeter (The name of the goddess for soil fertility has come in to force during 1924. Now *Demeter International* is an important certifying organisation in the world. Rudolf Steiner visited India and learned Indian farming techniques and integrated it with western techniques and formulated the biodynamic agriculture method.

Ewald Kunnemann, based on his farming knowledge published his book as three volumes on *Biological Soil Culture and Manure Economy*, respectively during the years 1931, 1932 and 1937. The book thought about the basics of natural farming.Similarly a book named L *agriculture biologique* by Claude Aubert of France is also a document on basics of natural farming.

The term organic farming was coined by Lord Northbourne in his book *Look to the Land* (written in 1939, published 1940). From his conception of "the farm as organism," he described a holistic, ecologically balanced approach to farming.

In 1939, influenced by Sir Albert Howard's work, Lady Eve Balfour launched the Haughley Experiment on farmland in England. It was the first scientific, side-by-side comparison of organic and conventional farming. Four years later, she published *The Living Soil*, based on the initial findings of the Haughley Experiment. Widely read, it led to the formation of a key international organic advocacy group, the Soil Association.

In Japan, Masanobu Fukuoka, a microbiologist working in soil science and plant pathology, began to doubt the modern agricultural movement. In 1937, he quit his job as a research scientist, returned to his family's farm in 1938, and devoted the next 60 years to developing a radical no-till organic method for growing grain and many other crops, now known as Nature Farming (Natural Farming), 'do–nothing' farming or Fukuoka farming.

Era of Development Post-World War II (1970–1990)

The research and practice of organic agriculture expanded worldwide after the 1960s.

In particular, the expansion and dual polarity of organic agriculture started with the oil crisis of 1973 and the growing sensitivity to agro-ecological issues. This was a time of new ideas, significant sociological transformations, protest movements and the proliferation of alternative life styles. The new thoughts in terms of using natural resources rationally, protecting the environment, realizing low input and high efficiency, ensuring food security, returning to the earth and maintaining a sustainable development of agriculture, such as organic, organic-biological, biodynamic ecological, and natural agriculture were remarkably developed in their concepts, research and practical activities.

In 1962, Rachel Carson, a prominent scientist and naturalist, published "*Silent Spring*", chronicling the effects of DDT and other pesticides on the environment. A bestseller in many countries, including the US, and widely read around the world, *Silent Spring* is widely considered as being a key factor in the US government's 1972 banning of DDT.

The book and its author are often credited with launching the worldwide environmental movement.

In the 1970s, global movements concerned with pollution and the environment increased their focus on organic farming. As the distinction between organic and conventional food became clearer, one goal of the organic movement was to encourage consumption of locally grown food, which was promoted through slogans like "Know Your Farmer, Know Your Food".

In 1972, the International Federation of Organic Agriculture Movements (IFOAM) was founded in Versailles, France and dedicated to the diffusion and exchange of information on the principles and practices of organic agriculture of all schools and across national and linguistic boundaries.

In 1975, Fukuoka released his first book, *The One-Straw Revolution*, with a strong impact in certain areas of the agricultural world. His approach to small-scale grain production emphasized a meticulous balance of the local farming ecosystem, and a minimum of human interference and labor.

In the U.S. during the 1970s and 1980s, J.I. Rodale and his Rodale Press (now Rodale, Inc.) were primary leaders in getting Americans to think about the side effects of non-organic methods and the advantages of organic ones. The press's books offered how-to information and advice to Americans interested in trying organic gardening and farming.

In the 1980s, around the world, farming and consumer groups began seriously pressuring for government regulation of organic production. This led to legislation and certification standards being enacted through the 1990s and to date.

Since the early 1990s, the retail market for organic farming in developed economies has been growing by about 20% annually due to increasing consumer demand. Concern for the quality and safety of food, and the potential for environmental damage from conventional agriculture, are apparently responsible for this trend.

Era of Growth – Twenty first century

The organic farming worldwide entered a new stage of growth in the 1990s. The trade organizations for organic products were founded, organic farming regulations were implemented and organic farming movement was promoted by both governmental and non governmental organizations.

The International federation of Organic Agriculture Movements (IFOAM) and the Food and Agriculture Organization of the United Nations (FAO) set out the guidelines *for the Production, Processing, Labeling and Marketing of Organically Produced Foods* in 1999. This guidelines are of importance to international harmonization of the organic farming standards.

Agribusiness is also changing the rules of the organic market. The rise of organic farming was driven by small, independent producers and by consumers. In recent years, explosive organic market growth has encouraged the participation of agribusiness interests. As the volume and variety of "organic" products increases, the viability of the small-scale organic farm is at risk, and the meaning of organic farming as an agricultural method is ever more easily confused with the related but separate areas of organic food and organic certification.

In Havana, Cuba, a unique situation has made organic food production a necessity. Since the collapse of the Soviet Union in 1989 and its economic support, Cuba has had to produce food in creative ways like instituting the world's only state-supported infrastructure to support urban food production called organopónicos, the city is able to provide an ever increasing amount of its produce organically.

Organic agriculture is holistic production management systems which promotes and enhances agro-ecosystem health, including biodiversity, biological cycles, and soil biological activity. It emphasizes the use of management practices in preference to the use of farm inputs, taking into account that regional conditions require locally adapted systems. This is accomplished by using, where possible, cultural, biological and mechanical methods, as opposed to using synthetic materials, to fulfils any specific function within the system terms, such as Organic, Biological, Biodynamic, and Ecological are recognized as organic farming in the EU regulations. Although organic agriculture is one among the broad spectrum of methodologies which are based on the specific and precise standards with different names such as organic, biological, organic-biological, bio-dynamic, natural and ecological agriculture, there are some common followed principles in the organic agriculture. These principles are summarized as follows:

- Maintain long-term soil fertility though biological mechanism.
- Recycle wastes of plant and animal origin in order to return nutrients to the land, thus minimizing the use of external inputs outside systems, and keep the nutrients cycle within the system.
- Prohibit the use of synthetic materials, such as pesticides, fertilisers, chemical ingredients and additives.
- Using natural mechanism and rely on renewable resources to protect the natural resources.
- Raise animals in restricted areas and guarantee the welfare of the animals.
- Adapt local environment and diversified organization.

The rapid growth of organic farming at global scale started during the end part of twentieth century, several trade organizations were founded, regulations were implemented and movements were promoted by both governmental and nongovernmental organizations. This led to rapid development of organic farming with co-ordinate and rational approach.

World Scenario of organic farming

According to the latest FiBL survey on certified organic agriculture worldwide, as of the end of 2017, data on organic agriculture was available from 179 countries (172 in 2014).

There were 50.9 million hectares of organic agricultural land in 2015, including inconversion areas. The regions with the largest areas of organic agricultural land are Oceania (22.8 million hectares, which is almost 45 per cent of the world's organic agricultural land) and Europe (12.7 million hectares, 25 per cent). Latin America has 6.7 million hectares (13 per cent) followed by Asia (4 million hectares, 8 per cent), North America (3 million hectares, 6 per cent), and Africa (1.7 million hectares, 3 per cent). The countries with the most organic agricultural land are Australia (22.7 million hectares), Argentina (3.1 million hectares), and the United States (2 million hectares).

Currently, one per cent of the world's agricultural land is organic. The highest organic shares of the total agricultural land, by region, are in Oceania (5.4 per cent) and in Europe (2.5 per cent). In the European Union, 6.2 per cent of the farmland is organic. However, some countries reach far higher shares: Liechtenstein (30.2 per cent) and Austria (21.3 per cent). In eleven countries, 10 per cent of the agricultural land or more is organic.

It was reported that there were almost 6.5 million hectares more of organic agricultural land in 2015 than in 2014. This is mainly because 4.4 million additional hectares were reported from Australia. However, many other countries reported an important increase thus contributing to the global growth, such as the United States (30 per cent increase) and India (64 per cent increase), both with an additional 0.5 million hectares, and Spain and France, both with an additional 0.3 million hectares. There has been an increase in organic agricultural land in all regions with the exception of Latin America; in Europe, the area grew by almost 1 million hectares (8.2 per cent increase).

In Africa, the area grew by almost 33.5 per cent or an additional 0.4 million hectares; in Asia, the area grew by 11 per cent or almost 0.4 million hectares, and in North America by more than 21 per cent or over 0.5 million additional hectares. Only in Latin America did the area of organic land decrease, mainly due to a decrease of almost 300'000 hectares in organic grazing areas in the Falkland Islands (Malvinas). A major relative increase of organic agricultural land was noted in many African countries, such as Kenya, Madagascar, Zimbabwe, and Côte d'Ivoire.

Apart from land dedicated to organic agriculture, there are further areas of organic land dedicated to other activities, most of these being areas of wild collection and beekeeping. Other areas include aquaculture, forests, and grazing areas on non-agricultural land. The areas of non-agricultural land constitute more than 39.7 million hectares.

Organic agricultural land

Currently, 50.9 million hectares are under organic agricultural management worldwide (end of 2015 for most data).[1]
The region with the most organic agricultural land is Oceania, with 22.8 million hectares followed by Europe with 12.7 million hectares, Latin America (6.7 million hectares), Asia (almost 4 million hectares), North America (almost 3 million hectares), and Africa (1.7 million hectares).

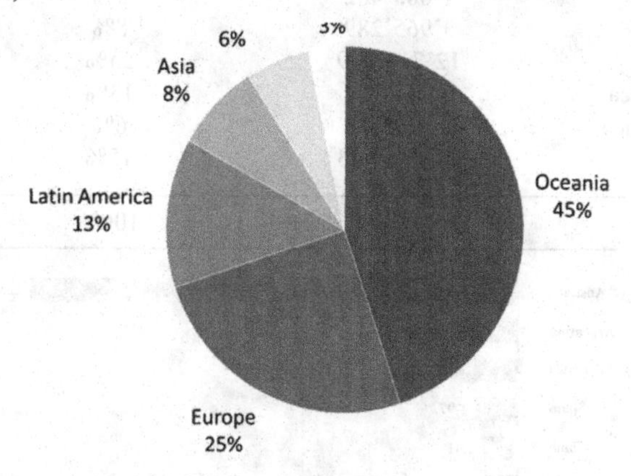

Fig. 1. World: Distribution of organic agricultural land by region 2015
Source: FIBL Survey 2017

Oceania has 45 per cent of the global organic agricultural land. Europe, a region that had a very constant growth of organic land over the years, has a quarter of the world's organic agricultural land followed by Latin America with 13 per cent (Table 1, Figure 1).

Australia, which experienced a major growth of organic land in 2015 (+4.4 million hectares), is the country with the most organic agricultural land; it is estimated that 97 per cent of the farmland are extensive grazing areas. Argentina is second followed by the United States in third place (Figure 2). The 10 countries with the largest organic agricultural areas have a combined total of 37.8 million hectares and constitute almost three-quarters of the world's organic agricultural land.

Apart from the organic agricultural land, there are further organic areas such as wild collection areas. These areas constitute more than 39.7 million hectares.

Table 1: World: Organic agricultural land (including in-conversion areas) and regions' shares of the global organic agricultural land 2015

Region	Organic agricultural land [hectares]	Regions' shares of the global organic agricultural land
Africa	1'683'482	3%
Asia	3'965'289	8%
Europe	12'716'969	25%
Latin America	6'744'722	13%
North America	2'973'886	6%
Oceania	22'838'513	45%
Total*	50'919'006	100%

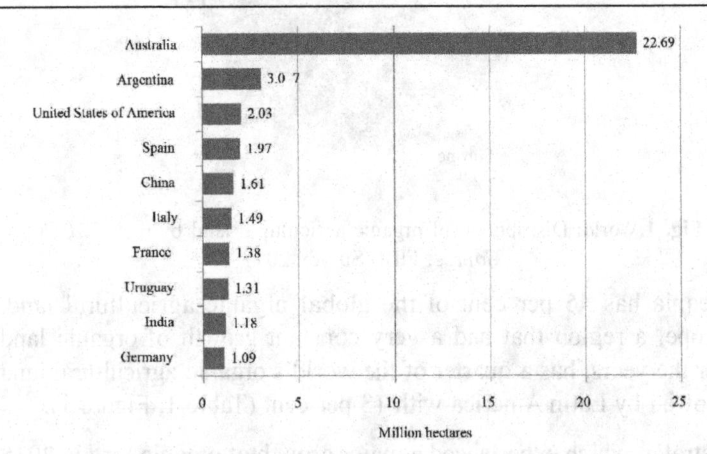

Fig. 2: World: The ten countries with the largest areas of organic agricultural land (FiBL Survey, 2017)

Table 2: Selected key crop groups and crops in organic agriculture 2015 (overview): Land under organic management (including conversion areas)

Crop	Africa [ha]	Asia [ha]	Europe [ha]	Latin America [ha]	North America [ha]	Oceania [ha]	Total [ha]
Cereals	72'361	900'352	2'232'921	124'849	558'870		3'889'353
Citrus fruit	6'586	7'293	42'520	10'383	4'017		70'798
Cocoa	110'067	2'332		187'242		2'765	302'406
Coffee	303'167	110'488		476'909		13'314	903'878
Dry pulses	15'988	18'554	328'870	6'666	38'343		408'421
Fruit, temperate	644	120'957	141'517	5'239	19'146	1'000	288'502
Fruit, tropical and subtropical	154'237	40'534	26'455	119'766		33'778	374'769
Grapes	1'538	16'745	292'753	7'224	12'623	2'022	332'905
Oilseeds	155'899	637'581	298'856	42'337	101'105		1'235'778
Olives	128'297	7'739	532'083	3'913			672'033
Vegetables	7'766	53'945	157'964	17'950	115'951		353'577

Global market

Global retail sales of organic food and drink reached 81.6 billion US dollars in 2015 (8160 crore USD) according to Organic Monitor, expanding about ten per cent compared to the previous year. North America and Europe generate the most organic product sales (90 per cent of organic food and drink sales). However, their global share of organic food sales is decreasing slightly as regional markets take root in Asia, Latin America, and Africa. Many of the organic crops grown in those regions are destined for exports. The global market for organic food and drink has expanded over almost four-fold between 2000 (18 billion US dollars) and 2015, and Organic Monitor projects growth to continue. However, there are a number of challenges: demand concentration in Europe and North America, the fact that in most countries, only a small consumer base is responsible for most organic food purchases, the challenge of marketing organic food according to consumer preferences in the various countries, and the concern about supply. Looking forward, positive growth in the organic products market is expected to continue in the coming years.

In 2015, the countries with the largest organic markets were the United States (35.8 billion euros), Germany (8.6 billion euros), and France (5.5 billion euros). The largest single market was the United States (approximately 47 per cent of the global market), followed by the European Union (27.1 billion euros, 35 per cent), and China (4.7 billion euros, 6 per cent). The highest per-capita consumption with more than 170 euros was found in Switzerland, Denmark, Luxembourg, and Sweden. The highest organic market shares were reached in Denmark (8.4 per cent), Switzerland (7.7 per cent), and Luxembourg (7.5 per cent).

A comparison of the global organic and Fairtrade market is provided by Lernoud and Willer on page 143. According to Fairtrade International, global Fairtrade sales reached 7.3 billion euros 2015. About 90 per cent of the sales of organic and Fairtrade products are in Europe and North America. For organic, North America is the largest market with over 50 per cent of the global organic market, while for Fairtrade products, Europe represents almost 80 per cent of Fairtrade retail sales.

Africa

There were almost 1.7 million hectares of certified organic agricultural land in Africa in 2015, which constitutes three per cent of the world's organic agricultural land. Comparing with 2014, Africa reported an increase of over 400'000 hectares, a 33 per cent increase and the largest growth since 2008. There were more than 700'000 producers. The United Republic of Tanzania was the country with the largest organic area (with almost 270'000 hectares), and Ethiopia was the country with the largest number of organic producers (more than 200'000). The country with the highest share of organic agricultural land was the island state Sao Tome and Principe, with 13.8 per cent of its agricultural area being organic. The majority of certified organic produce in Africa is destined for export markets. Key crops are coffee, olives, nuts, cocoa, oilseeds, and cotton. In Africa, only Morocco and Tunisia have an organic regulation; seven countries are drafting one, and eleven countries have a national standard but not a national legislation.

The policy brief of United Nations Conference on Trade and Development (UNCTAD) "*Financing Organic Agriculture in Africa: Exploring the Issues*" (UNCTAD 2016) was published as a support to elevate financing of the sector in the continent. According to this report, organic agriculture is a rapidly growing sector in Africa, with strong links to economic and sociocultural development. Organic conferences in Eastern, Western, Central and Southern Africa have become a success, and the most recent Eastern Africa conference was held in 2016, in Entebbe, Uganda. These conferences marked significant milestones for mainstreaming organic agriculture in policies, strategies, and programmes.

In Kenya, the compilation of organic sector data for 2015 showed an impressive growth compared to the 2011 figures. The demand for organic food has continued to grow with the urban rich, providing huge market opportunities, as shown by a recent study. If the trend continues towards 2016/2017, the projection is that more farmers are likely to convert to organic farming as the demand for organic products such as coffee and tea will be unmet and on the rise.

Asia

The total area dedicated to organic agriculture in Asia was almost 4 million hectares in 2015. There were more than 0.8 million producers; most of these were in India. The leading countries by area were China (1.6 million hectares) and India (almost 1.2 million hectares); Timor-Leste had the highest proportion of organic agricultural land (6.6 per cent). Nineteen countries have regulations on organic agriculture, and five countries are in the process of drafting one.

Asia's share of organic food sales continues to rise. China has the largest market in the region. The spate of food scares in Asia has been a major driver of organic food sales (see the chapter by Amarjit Sahota, page 138). India, aside from being an exporter, has a growing domestic market for organic products. The rise in the income of the urban middle class has fuelled an increase in the demand for organic food, particularly in the cities.

Many countries now support organic agriculture such as China, which signed the first bilateral organic certification agreement with New Zealand. Furthermore, the Chinese central government has now also decided to incorporate the organic industry into its "National Plan for the Construction of Ecological Civilization." National organic policies have been approved in Bangladesh and Kyrgyzstan in 2016, and in South Korea checkoff funds are now mandatory. Participatory Guarantee Systems (PGS) reported a steady growth in Asia, and some governments have accepted PGS as an alternative form of certification for organic products.

Understanding the importance of the role of local governments in the adoption and implementation of organic agriculture practices, IFOAM Asia initiated the "Asian Local Governments for Organic Agriculture". An annual summit brings together representatives from both the public and private sectors to discuss issues related to the development of organic agriculture in Asia. The second Organic Asia Congress will be held in May 2017 in China. For more information including country reports, see the chapter from IFOAM Asia.

Europe

As of the end of 2015, 12.7 million hectares of agricultural land in Europe (European Union 11.2 million hectares) were managed organically by almost 350'000 producers (European Union almost 270'000). In Europe, 2.5 per cent of the agricultural area was organic (European Union: 6.2 per cent). Twenty-five per cent of the world's organic land is in Europe. Organic farmland has increased by approximately one million hectares compared to 2014. The countries with the largest organic agricultural areas were Spain (almost 2 million hectares), Italy (1.5 million hectares), and France (1.4 million hectares). In nine countries at least 10 per cent of the farmland is organic: Liechtenstein has the lead (30.2 per cent), followed by Austria (21.3 per cent) and Sweden (16.9 per cent). Retail sales of organic products totalled approximately 29.8 billion euros in 2015 (European Union: 27.1 billion euros), an increase of 13 per cent over 2014. The largest market for organic products in 2015 was Germany, with retail sales of 8.6 billion euros, followed by France (5.5 billion euros), and the UK (2.6 billion euros) (see the article by Willer *et al.*, page 207). Despite the dynamic market growth, current trends indicate that production in Europe is not moving at the same speed, which presents several challenges for the future development of organic in Europe.

In Europe, all countries have an organic regulation or are drafting one. The revision of the European Union (EU) regulation on organic farming, which applies in all EU countries, was an important topic in 2016; twelve months after the start of trilogue negotiations on the European Commission's legislative proposal between the European Parliament, Agriculture Council, and European Commission, talks remained deadlocked at the end of 2016. Positions amongst the EU Institutions and the member states themselves continue to diverge on key topics. The EU Common Agricultural Policy (CAP) and similar programmes in other countries remain a key policy for the development of agriculture in Europe, including organic farming. Under the current CAP for the period 2014-2020 organic farming is supported by Pillar 1 (direct payments) and Pillar 2 (Rural Development Programmes). On the research end, in 2016, the European Technology Platform for Organic Food and Farming

Research (TP Organics) published priority topics for the Work Programme 2018/2020 of Horizon 2020, the current research framework programme of the European Union.

Latin America and the Caribbean

In Latin America, almost 460'000 producers managed 6.7 million hectares of agricultural land organically in 2015. This constituted 13 per cent of the world's organic land and almost one per cent of the region's agricultural land. The leading countries were Argentina (3.1 million hectares), Uruguay (1.3 million hectares), and Brazil (0.75 million hectares, 2014). The highest shares of organic agricultural land were in the Falkland Islands/Malvinas (12.5 per cent), Uruguay (9 per cent), and French Guiana (9 per cent). Many Latin American countries remain important exporters of organic products such as bananas, cocoa, and coffee; in countries such as Argentina and Uruguay, temperate fruit and meat are key export commodities. Twenty-three countries in this region have an organic regulation or are drafting one. In May 2016, the European Union and Chile concluded negotiations of an agreement on trade in organic products to mutually recognize the equivalence of their organic production rules and control systems.

Organic production in the region largely depends on cooperation between smallholders, especially in coffee, cacao, banana, mango, Andean grains, and ginger value chains. The capacity of Latin American countries to develop their organic sectors can be improved with incentives and governmental support, and local governments are taking the lead in several national and decentralized initiatives (for instance Argentina) including support for Participatory Guarantee Systems PGS (for instance Peru).

North America

In North America, almost 3 million hectares of farmland were managed organically in 2015. Of these, 2 million were in the United States and 0.9 million in Canada, representing 0.7 per cent of the total agricultural area in the region and 6 per cent of the world's organic agricultural land.

The booming organic industry in the United States continues to set new records, with total organic product sales hitting 43.3 billion US dollars[2] by the end of 2015, up 11 per cent from the 2014 record level and outstripping the overall food market's growth rate of 3 per cent, according to the Organic Trade Association. Of the 43.3 billion dollars in total organic sales, 39.7 billion dollars were organic food sales. The United States Department of Agriculture (USDA) in mid-January 2017 officially proposed a nationwide research and promotion check-off program for the organic industry to comment on and ultimately vote on. The USDA proposal estimates the organic check-off could raise over 30 million US dollars a year to spend on research to make farmers successful, technical services to accelerate the adoption of organic practices, and consumer education and promotion of the organic brand. In 2016, the Organic Farming Research Foundation released a report analyzing organic farming and food research in the United States, and the report found that about three-quarters of the funding supported research on organic crop production, with the remainder going to livestock, crop-livestock systems, and general topics.

Organic products continue to enjoy a robust demand in Canada. The domestic consumer demand is estimated at 4.7 billion Canadian dollars[3] in retail sales in 2015, a 1.2 billion Canadian dollar increase from 2012. In the past decade, Canada's organic market has been experiencing a double-digit annual growth rate, and growth is expected to continue. Canada is one of the few countries that tracks imported organic products using Harmonized System (HS) codes (limited mainly to imported fresh fruit and vegetables, coffee and tea, and dairy products). According to this data, in 2015, Canada imported 652 million Canadian dollars' worth of the 65 tracked organic products, representing a 37 per cent increase from 2012.

Oceania

This region includes Australia, New Zealand, and the Pacific Island states. Altogether, there were over 22'000 producers, managing 22.8 million hectares. This constituted 5.4 per cent of the agricultural land in the region and 45 per cent of the world's organic land. More than 99 per cent of the organic land in the region is in Australia (22.7

million hectares, 97 per cent of which is estimated to be extensive grazing land), followed by New Zealand (more than 74'000 hectares), and Samoa (almost 28'000 hectares). The highest organic shares of all agricultural land were in Samoa (9.8 per cent), followed by Tonga (8 per cent), Australia (5.6 per cent), the Solomon Islands (5.2 per cent), and Vanuatu (5.1 per cent). Growth in the organic industry in Australia, New Zealand, and the Pacific Islands has been strongly influenced by a rapidly growing overseas demand; domestic sales are also growing. In Australia, the domestic market was valued at 1.3 billion Australian dollars (data from 2014[4])and in New Zealand at 197 million New Zealand dollars in 2015.

The area of land in Australia under certified organic management continues to grow; the majority of the organic area is used for beef cattle production in the semi-arid rangelands, where individual pastoral operations typically occupy tens of thousands of hectares each. The regulatory framework for organic certification in Australia has remained stable with little change in 2016. However, the organic industry and Australian Government continue to respond to global organic developments through review of the National Standard for Organic and BioDynamic Produce (National Standard), which was revised in 2016. Most Australian shoppers – 59 per cent of all shoppers in 2016 – are aware that certification marks are used on organic products as a guarantee of authenticity.

In 2016, the value of organic agriculture as a development tool was recognized by the Pacific Communities governing body, the Council of Regional Governments and Administrations, which consists of the ministries of foreign affairs and trade of the 26 Pacific Community member states. Important developments in 2016 included the Pacific Organic Tourism and Hospitality Standard, which was developed with the assistance of the European Union Pacific Agriculture Policy Project during 2016, and an organic policy toolkit for government policy- and decision-makers. Most of the organically certified products from the region are for export; however, there are indications of growing local markets.

Indian scenario

The growth of organic agriculture in India has three dimensions and is being adopted by farmers for different reasons. First category of organic farmers are those which are situated in no-input or low-input use zones; for them organic is a way of life and they are doing it as a tradition. Second category of farmers are those which have recently adopted the organic in the wake of ill - effects of modern agriculture, may be in the form of reduced soil fertility, food toxicity or increasing cost and diminishing returns. The third category comprise of farmers and enterprises, which have systematically adopted the commercial organic agriculture to capture emerging market opportunities and premium prices. While majority of the farmers in first category are traditionally (or by default) organic they are not certified; second category farmers comprise of both certified and uncertified, but majority of third category farmers are certified and are the commercial organic farmers.

During the past decade, there has been significant growth in the area of organic agriculture. There has been almost a three-fold increase, from 528'171 hectares in 2007-2008 to 1.18 million hectares of cultivable land in 2014-15. The data does include the 3.71 million hectares of forest land and wild areas for collection of forest produce.

The significant growth is attributed mainly to conducive policies that have led to an increase in areas under third-party certification and has promoted Participatory Guarantee Systems (PGS). Some of the pioneering civil society organizations involved in facilitating PGS have influenced government policies in favour of PGS. India is among the few countries of the world where PGS is recognized and promoted by the government.

In addition to the area certified as organic, there are vast tracts of land that are traditionally organic but not certified as such. For instance, the State of Sikkim with an area of 70'000 hectares has been declared as an organic state with regulations that prohibit the use of chemical fertilizers and pesticides. There are still other states that are almost entirely organic like the State of Nagaland with an area of 1.6 million hectares. Besides, there are several rain-fed farms mostly in the central part of the country that are organic by nature.

The inclusion of such farms that are traditionally organic into formal certification systems will significantly increase the organic area under certification, and more certified organic produce will be available in the markets.

Besides being an exporter, India also has a growing domestic market for organic products. The rise in the incomes of the urban middle class has fuelled increased demand for organic foods, particularly in the cities.

Of the total 23 lakh organic producers of the world, India ranks first and has 5.85 lakh producers followed by 2.03 lakhs in Ethiopia and 2.0 lakh in Mexico.

As per a study by ASSOCHAM India, the organic food turnover is growing at about 25 per cent annually and is expected to reach 1.36 billion US dollars by 2020 from 0.36 billion US dollars in 2014.

Organic farming as a means to sustainable agriculture has benefited farmers. The certified cultivated area under organic farming has grown from 4.55 lakh ha in 2009-10 to 7.23 lakh ha in 2013-14, with around 6 lakh farmers practising it. But, still, the total area under organic farming is insignificant compared to the net sown area of 140 million hectares. In terms of exports also, exports of organic food at about 1.6 lakh tonnes and at an estimated value of USD 220 million is less than 1 per cent of global exports. Against this backdrop, to provide a major fillip to organic farming in India, the existing components of organic farming under the NMSA have been put together under a new programme called "Paramparagat Krishi Vikas Yojana". The programme envisages development of 10,000 organic clusters and provides chemical-free inputs to farmers and increase the certified area by 5 lakh hectare within a period of 3 years. Under this, every farmer in a cluster will be provided an assistance of Rs. 50,000 per hectare in 3 years towards conversion to and adoption of organic farming and towards market assistance.

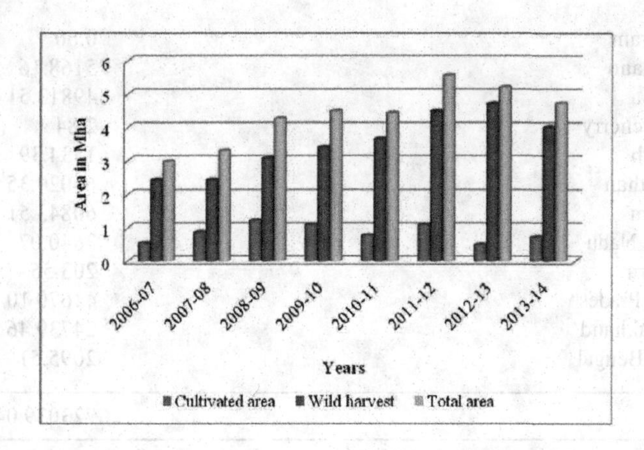

Fig. 3: Cultivated, Wild and Total Area under Organic Certification in India

Table 3. State-wise farm area under organic certification (excluding forest area) during 2013-14

Name of States	Area (ha)
Andaman and Nicobar Islands	321.28
Andhra Pradesh	12325.03
Arunachal Pradesh	71.49
Assam	2828.26
Bihar	180.60
Chhattisgarh	4113.25
Delhi	0.83
Goa	12853.94
Gujarat	46863.89
Haryana	3835.78
Himachal Pradesh	4686.05
Jammu & Kashmir	10035.38
Jharkhand	762.30
Karnataka	30716.21
Kerala	15020.23
Lakshadweep	898.91
Madhya Pradesh	232887.36
Maharashtra	85536.66
Manipur	0.00
Meghalaya	373.13

(Contd.)

Mizoram	0.00
Nagaland	5168.16
Odisha	49813.51
Pondicherry	2.84
Punjab	1534.39
Rajasthan	66020.35
Sikkim	60843.51
Tamil Nadu	3640.07
Tripura	203.56
Uttar Pradesh	44670.10
Uttarakhand	24739.46
West Bengal	2095.51
Total	723039.04

Tamil Nadu scenario

Organic farming in Tamil Nadu

Tamil Nadu is going organic, but rather slowly. There is a distinct movement in Tamil Nadu among the farmers, agriculture experts and scientists in favour of organic farming. In the year 2007-2008, Tamil Nadu Organic Certification Department (TNOCD) was established to carryout inspection and certification of organic production system in accordance with NPOP (National Programme for Organic Production). It also imparts free training to registered organic farmers on National Standards for Organic Production (NSOP) and Tamil Nadu Organic Certification Department Standards. NGOs, Tamil Nadu Agricultural University, State Department of Agriculture and other government and private agencies have started advocating organic farming in major crops.

Table 4. Status of organic farming and organic farmers in Tamil Nadu (TNOCD, 2017)

Year	Individual farmers		Group of farmers		Corporate Sector	Total	
	Area (ac)	No. of farmers	Area (ac)	No. of farmers	Area (ac)	Area (ac)	No. of farmers
2007-08	3185.05	33	1059.21	232	1000.00	5244.00	466
2008-09	5270.84	98	19609.70	7575	167.07	25048.00	7673
2009-10	6428.43	150	21306.00	9516	484.46	28219.00	9666
2010-11	8204.27	247	19487.50	7945	422.86	28114.59	8192
2011-12	7972.06	209	20392.20	6127	698.00	29062.40	6336
2012-13	7042.30	220	15568.79	5900	266.43	22877.52	6120
2013-14	8300.87	230	19243.16	4394	474.63	28018.66	4624
2014-15	8190.59	276	20342.55	4730	1144.23	29677.37	5006
2015-16	9198.97	329	19597.94	4070	929.90	29719.81	4412
2016-17	-	348	-	4370	-	30910.53	4718

At present among the districts of Tamil Nadu, Cuddalore has the highest acreage under certified organic agriculture (9524) followed by Salem (3645), Nagapattinam (2564), Villupuram (1392), Dindigul (1369), Coimbatore (1349), Tuticorin (1327), Thiruvallur (1204) Tiruppur (1150) and Erode (1034).(Footnotes)

Chapter 3

Government Initiatives and Research Institutes

The Tenth Five Year Plan (GoI 2002), for 2002 through 2007, has put emphasis on natural resource management through rainwater harvesting, groundwater recharging measures and controlling groundwater exploitation, watershed development, treatment of waterlogged areas. With regard to application of agricultural inputs like fertilizer and pesticides, the Plan stated that factors such as imbalanced use of nitrogenous (N), phosphatic (P) and postassic (K) fertilisers, increased deficiency of micronutrients and decreased soil organic carbon would be addressed through a holistic agri-environmental approach stressing Integrated Plant Nutrient and Pest Management. Further, the Tenth Plan document recognizes organic farming as a 'thrust area' in the sustainable use and management of resources in agriculture.

The trajectory of Indian agriculture and its associated environmental problems has brought about recognition that future agricultural growth and productivity will have to occur simultaneously with environmental sustainability. The environmental challenges, especially in terms of land degradation and groundwater depletion, water logging and excessive use of chemical inputs are posing problems for the future of Indian agriculture. To address the problems, policies have laid emphasis on promoting sustainable agriculture including organic farming. Differential approaches and policy instruments, however, will be required to address these problems. The shift from input-intensive to sustainable, particularly organic farming is a difficult task as it involves a number of policy measures dealing with a variety of issues ranging from the transfer of information and technology to the development of markets. Another difficult task, and perhaps more difficult, relates to marginal and small farmers – which comprise a substantial part of Indian agriculture. Although these marginal and small farmers have been considered organic by 'default', severe resource constraints make a shift to the modern sense of organic farming prohibitive.

The Indian government has been undertaking measures to promote organic farming with the aim to improve soil fertility and help to double the farmers' incomes by the year 2022. The Prime Minister had visited Sikkim—which is India's first organic state—and encouraged other states to replicate the "Sikkim model". Some of the policy initiatives to promote organic farming and exports include development of an organic regulation for exports by the Agricultural and Processed Food Products Export Development Authority (APEDA), removal of quantitative restriction on organic food exports, providing subsidies to farmers under the Paramparagat Krishi Vikas Yojana (PKVY) in partnership with the state governments, and other schemes such as the Mission Organic Value Chain Development for North Eastern Region. Despite these initiatives, a recent survey-based study covering 418 organic farmers across different states of India suggests that a move to organic farming methods may not be that easy and organic farmers are not getting the expected premium price for their produce.

Government Initiatives to Promote Organic Farming

i) National Project on Organic Farming

The National Project on Organic Farming (NPOF) is a Central Sector Scheme implemented during theTenth Five Year Plan with an outlayof Rs. 57.04 crore. The scheme was subsequently expanded in the Eleventh Five Year Plan with an outlay of Rs. 101 crore. The primary objective of the NPOF Scheme is to encourage the production of food organically, and promote manufacture and usage of organic and biological inputs, such as bio-fertilizers, organic manure, biopesticides and bio-control agents.

ii) Capital Investment Subsidy for Setting up of Organic InputsProduction

The NPOF provides financialassistance for fruits and vegetables waste compost units by providing for 33 per cent of the capital cost of the project, subject to a ceiling of Rs. 63 lakh. Further, NPOF provides subsidy for the construction of biofertilizer or bio pesticide production unit to an extent of 25 per cent of thecapital cost of the project subject to aceiling of Rs. 40 lakh. The remainingcost is envisaged as credit support from financial institutions and marginmoney. The subsidy is credit linkedand back-ended and mobilisedthrough NABARD.

iii) National Project on Management of Soil Health and Fertility (NPMSHF)

The National Project on Management of Soil Health and Fertility (NPMSF) was implemented during the Eleventh Five Year Plan period with an outlay of Rs. 429.85 crore, to promotethe balanced and judicious use of fertilizers and organic manure on soiltest basis. This Scheme provides financial assistance at Rs. 500 perha for promoting the use of organicmanure.

iv) Network Project on OrganicFarming by ICAR

The Network Project on Organic Farming initiated by the ICAR in the Tenth Five Year Plan at the Project Directorate for Farming Systems Research, Modipuram, Uttar Pradesh, involves developing package of practices for different crops andfarming systems under organicin different agro-ecologicalregions of the country. The project has been running at 13 centres including State Agricultural Universities (SAUs), spread across 12 States. The crops for which package of practices for organic farming have been developed include basmati rice, rain fed wheat, maize, red gram, chickpea, soybean, groundnut, mustard, isabgol, blackpepper, ginger, tomato, cabbage and cauliflower.

v) National Horticulture Mission

This is a Centrally Sponsored Scheme; launched in 2005-06, the Schemeaims at strengthening the growth of the horticulture sector comprising of fruits, vegetables, roots and tubercrops, mushroom, spices, flowers, aromatic plants, cashew and cocoa. NHM provides financial assistance forestablishing vermi compost units and HDPE vermi beds. Assistance is also being provided under the Mission for organic certification of Rs.5 lakh for agroup of farmers covering an area of 50 hectares.

vi) Rashtriya Krishi Vikas Yojna

Assistance for decentralized production and marketing of organic fertilizers is available under Rashtriya Krishi Vikas Yojna (RKVY) for projects formulated and approved by the State Level Sanctioning Committee.

vii) ICAR Contribution in Promoting Organic Farming

All India Network Project on Soil Biodiversity-Biofertilizers is implemented by Indian Council of Agricultural Research (ICAR) for R & D on biofertilizers. The ICAR has developed technologies to prepare various types of organic manures such as phosphocompost, vermi compost, municipal solid waste compost etc. Improved and efficient strains of biofertilizers specific to different crops and soil types are being developed under Network Project on biofertilizers. The financial assistance is provided on the basis of project proposals received from States including Maharashtra. Indian Council of Agricultural Research (ICAR) under Network Project on Organic Farming, with lead centre at Project Directorate for Farming Systems Research Modipuram is developing package of practices of different crops and cropping system under organic farming in different agro-ecological regions of the country.It is important to note that India exported agri-organic products of total volume of 160276.95 MT and realization was around Rs.1155.81 crores in year 2012-13.

viii) Paramparagat Krishi Vikas Yojana (PKVY)

The Ministry of Agriculture & Farmers Welfare is promoting Organic Farming as a sub-component under National Mission on Sustainable Agriculture (NMSA). Under the scheme, financial assistance is provided for setting up of mechanized Fruit and Vegetable market wastes, agro wastes compost units, setting up of liquid carrier-based biofertilizer biopesticide production units. In order to promote participatory certification of Organic Farming in a cluster approach, Paramparagat Krishi Vikas Yojana (PKVY) was formulated in year 2014-15. The various components of NMSA are: (a) adoption of organic farming through cluster approach under Participatory Guarantee System (PGS) certification, (b) support to PGS system for online data management and residue analysis, (c) training and demonstration on organic farming, (d) organic village adoption for manure management and biological nitrogen harvesting, have been clubbed together under PKVY.

Objectives

1. To launch eco-friendly concept of cultivation reducing the dependency on agro-chemicals and fertilizers.
2. To optimally utilize available natural resources for input production.
3. To create employment opportunities in the rural as well as urban sector.
4. To develop potential market for organic products.
5. To develop an organic village with a cluster of 50 farmers and 50 ha land with a concept of 50:50 for both agriculture and horticulture emphasizing mixed cropping instead of mono cropping.
6. To use the viable technology under IPM, INM and local indigenous traditional techniques for management of plant nutrition and plant protection.
7. Promotion of low-cost PGS certification system for organic produce.
8. Promotion of marketing and branding of organic produce through financial assistance for packaging, labelling, PGS logo and hologram.

The new components under PKVY have been designated as: (a) adoption of PGS certification through cluster approach; (b) adoption of organic village for manure management and biological nitrogen harvesting through cluster approach. The DAC&FW intends to enhance the area under organic farming to 5.0 lakh acres within a period of 3 years. It is targeted to cover 10000 clusters of farmers (about 50 farmers in each cluster) in three years with an expenditure of Rs 1,495 crore.

GoI Budget 2018-19 – Operation Green

Rs. 500 crore operation greens announced to address price volatility of perishable commodities like potato, tomato and onion and benefit both producers and consumers. To promote Farmer Producers Organisations, Village Producers Organizations (VPOs), processing facilities, agri-logistics and professional management.

Agencies and institutions related to organic farming

Table 5. International institutes for organic farming

1.	Agrecol
2.	Association of Ecological Producer Organizations in Bolivia (AOPEB)
3.	Agricultura Ecologica
4.	Agriculture and Organic Farming Group
5.	Alternative Farming Systems Information Center (AFSIC)
6.	Annadana Soil and Seed Savers Network
7.	Appropriate Technology Transfer for Rural Areas (ATTRA)
8.	Association Kokopelli
9.	Australia National Standards
10.	BioFach -The World Organic Trade Fair
11.	Details Argentinan Standards for Organic Production
12.	Biodynamic Farming and Gardening Association (BDA)
13.	CAB International
14.	Details California Certification Standards
15.	Details Codex Alimentarius Commission
16.	Canadian Organic Growers (COG)
17.	Canberra Organic Growers Society (COGS)
18.	Center for Integrated Agricultural Systems (CIAS)
19.	Centre National de Ressources en Agriculture Biologique (CNRAB) - French only
20.	CORE Organic

(Contd.)

21. Danish Institute of Agricultural Sciences
22. Danish Research Centre for Organic Farming (DARCOF)
23. Demeter International
24. Department of Organic Farming and Cropping Systems, University of Kassel
25. Details Ecolabels
26. Economic Research Service of US Department of Agriculture
27. Ecological Knowledge Portal: EcoPort
28. European Network for Scientific Research Coordination in Organic Farming (ENOF)
29. European Weed Research Society (EWRS)
30. Environmentally and Socially Responsible Horticulture Production and Trade
31. Details Eosta
32. Details Essay on traceability
33. Ethical Trading Initiative (ETI)
34. EU Organic Regulations
35. European Information System for Organic Markets
36. European Network for Scientific Research Coordination in Organic Farming (ENOF)
37. Details European Society for Agriculture and Food Ethics (EurSafe)
38. European Weed Research Society (EWRS)
39. FAOLEX
40. Farming solutions
41. Faculty of Organic Agricultural Sciences, University of Kassel
42. Fairtrade Labelling Organizations International (FLO)
43. Farming Solutions
44. Forest Stewardship Council (FSC)
45. Gateway to Farm Animal Welfare
46. German Institute for Tropical and Subtropical Agriculture (DITSL)
47. Green Trade Net
48. Harmonization and Equivalence in Organic Agriculture
49. Goetheanum Section on Agriculture
50. Grolink and Swedish International Development Cooperation Agency (SIDA)
51. Henry A. Wallace Centre for Agricultural and Environmental Policy at Winrock International
52. Indonesia Organic

(Contd.)

53. Institute for Biodynamic Research
54. Independent Organic Inspectors Association (IOIA)
55. Indonesia Organic
56. Information Network on Post-Harvest Operations (InPho)
57. Details Institute for Biodynamic Research
58. Instituto Biodinamico Certification Association (IBD)
59. Details Instituto Biodinamico Certification Association (IBD)
60. Intergovernmental Group on Citrus Fruit
61. International Organic Accreditation Service (IOAS)
62. International Research Association for Organic Food Quality and Health (FQH)
63. Details International Social and Environmental Accreditation and Labelling Alliance (ISEAL)
64. Japan Agricultural Standards for Organics
65. Leopold Center for Sustainable Agriculture
66. Louis Bolk Institute
67. Market Briefs
68. Institute of Organic Agriculture (University of Bonn, Germany)
69. International Society of Organic Agriculture Research (ISOFAR)
70. OCIA International Organic Crop Improvement Association
71. Leopold Center for Sustainable Agriculture
72. Louis Bolk Institute
73. Living Soil Association of Tasmania
74. Michael Fields Agricultural Institute
75. Multinational Exchange for Sustainable Agriculture (MESA)
76. National Organic Program
77. Network of Researchers Involved in Organic Crop Rotation Experiments
78. Natural and Organic Products Europe
79. Details Natural Resources and Ethical Trade (NRET)
80. Nature & Progrès Belgique
81. Online Information Service for Non-Chemical Pest Management in the Tropics (OISAT)
82. Organic Agriculture Centre of Canada (OACC)
83. Organic Agriculture Information Access
84. Organic Centre Wales
85. Organic Consumers Association (OCA)
86. Organic Farming Research Foundation (OFRF)

(Contd.)

87. Organic Fertilisers and Water-Retaining Products Database
88. Organic Materials Review Institute (OMRI)
89. Organic Monitor
90. Organic Research Database
91. Organic Trade Association (OTA)
92. Organic X Seeds
93. OrganicAgInfo
94. Organic Agriculture and Fair Trade in West Africa
95. Organic Food Development Center
96. Organic Food Federation
97. Organic Research Database
98. OrganicAgInfo
99. Permaculture International, Ltd. (PIL)
100. Promoting Organic and Sustainable Agriculture Development in Africa (POSADA)
101. Pesticide Action Network (PAN)
102. Scottish Agricultural College (SAC)
103. Sociedad Española de Agricultura Ecológica (SEAE)
104. Soil & Health Association of New Zealand
105. Soil Association
106. Sustainable Agriculture Research and Education (SARE)
107. The Center for Agroecology and Sustainable Food Systems (CASFS) Apprenticeship Programme
108. The Egyptian biodynamic association (EBDA)
109. The New Farm
110. The Research Institute of Organic Agriculture (FiBL)
111. The Veterinary Epidemiology and Economics Research Unit (VEERU)
112. United Nations Economic and Social Commission of Asia and the Pacific (ESCAP)
113. University of California Sustainable Agriculture Research and Education Program (UC-SAREP)
114. Tanzania Organic Agriculture movement
115. Rwanda Organic Agriculture Movement (ROAM)
116. Promoting Organic and Sustainable Agriculture Development in Africa (POSADA), Ruwanda

Chapter 4

Biodiversity and Organic Farming

The word biodiversity refers to the variety of living organisms (flora and fauna). Biodiversity or Biological diversity is defined as the variability among all living organisms from all sources, including terrestrial, marine and other aquatic ecosystems and ecological complexes of which they are part. Wilson, 1988 defined 'Biological diversity' or 'biodiversity' as that part of nature which includes the differences in genes among the individuals of a species, the variety and richness of all the plant and animal species at different scales in space i.e. local, regional, country wise and global, andvarious types of ecosystems- both terrestrial and aquatic-within a defined area.

The diversity of agro-ecosystems

Biological diversity deals with the degree of nature's variety in the biosphere. This variety can be observed at three levels *i.e.,* genetic, species and ecosystem.

a) Genetic Diversity

Genetic diversity refers to the variation at the level of individual genes. Tremendous amount of genetic diversity exists within individual species. This genetic variability is responsible for the different characters in species. Genetic diversity is the raw material from which new species arise through evolution. Today, the genetic diversity is made use to breed new crop varieties, disease resistant crops.

Organic producers, on the other hand, are looking for varieties that are suited to their local climatic and soil conditions and are not susceptible to disease and pest attack. Research has shown that in general these characteristics are more likely to be found in the older native cultivars. For example, conventional wheat grown in an organic system will have reduced protein content, whereas the selection of a variety suitable for the particular growing conditions will provide higher protein, and thus higher baking quality wheat to be grown. Research has also indicated that yields and disease resistance are better in native cultivars as opposed to modern varieties for vegetables (tomato, cucumber and melon) grown in an organic system.

b) Species diversity

The number of species of plants and animals that are present in a region constitutes its species diversity. This diversity is seen both in natural ecosystem and in agricultural ecosystem. Some areas are richer in species than others. For example, natural undisturbed tropical forests have much greater species richness than monoculture plantations developed by the forest department for timber products. A natural forest ecosystem provides large number of non-timber forest products that local people depend on such as fruits, fuel, wood, fodder, fiber, gum, resin and medicines. Timber plantations do not provide the large variety of goods that are essential for local consumption. Modern intensive agro ecosystem has a relatively lower density of crops than traditional agro-pastoral farming systems, where multiple crops were planted.

Areas that are rich in species diversity are called 'hotspots' of diversity and the countries with highest species richness or have a relatively large proportion of these hot spots of diversity are referred to as 'megadiversity nations'. India is among the world's 15 nations that are exceptionally rich in species diversity. The earth's biodiversity is distributed in specific ecological regions. There are over a thousand major eco-regions in the world. Of these, 200 are said to be the richest, rarest and most distinctive natural areas. These areas are referred to as the Global 200. It has been estimated that 50,000 endemic plants which comprise 20% of global plant life, probably occur in only 25 *'hotspots'* in the world. These hotspots harbor many rare and endangered species. Two criteria help in defining hotspots namely rich endemism and the degree of threat. To qualify as hotspots an area must contain at least 0.5 per cent or 1500 of the worlds 3, 00,000 plants species as endemics.

Organic farms are also more likely to have higher agro-biodiversity with greater crop rotation diversity, number of cultivated crops and grassland composition.

c) Ecosystem diversity

An ecosystem is made up of the organisms of a particular habitat, such as a farm or forest, together with the physical landscape in which they live. There are a large variety of different ecosystem on earth, each having their own complement of distinctive inter linked species based on differences in the habitat. Ecosystem diversity can be described for a specific geographical region or a political entity such as a country, a state or a taluk. Distinctive ecosystems include landscapes like forests, grasslands, deserts, mountains *etc* as well as aquatic ecosystems like rivers, lakes and seas. Each region also has man- modified areas such as farmland or grazing pastures. It refers to the variation in the structure and functions of the ecosystem. It describes the number of niches, trophic levels and various ecological processes that sustain energy flow, flood webs and the recycling of nutrients. It has focus on various biotic interactions and the role and functions of **keystone species** (species determining the ability of large number of other species to persist in the community).

Methods of measuring Biodiversity

There are three perspectives measuring of diversity at the level of community.

These are (i) Alpha diversity, (ii) beta diversity and (iii) gamma diversity. Community diversity refers to the variations in the biological communities in which species live.

(i) Alpha diversity indicates diversity within the community. It refers to the diversity of organisms sharing the same community or habitat. A combination of species richness and equitability / evenness is used to represent diversity within a community or habitat.

(ii) **Beta diversity** indicates diversity between communities. Species frequently change when habitat or community changes. There are differences in species composition of communities along environmental gradients, *e.g,* altitudinal gradient, moisture gradient, etc. the higher heterogeneity in the habitats ina region or greater dissimilarity between communities exhibit higher beta diversity.

(iii) **Gamma diversity** refers to the diversity of the habitats over the total land scope orgeographical area. The sum of alpha and beta diversities of the ecosystems is anexpression of the biodiversity of landscape, which is considered as Gamma Diversity.

Higher diversity at community level provides stability and higher productivity. Intemperate grasslands, it has been observed that diverse communities are functionally more productive and stable, even under environmental stresses such as prolonged dry conditions.

Important uses of biodiversity

Environmental services from species and ecosystems are essential at global, regional and locallevels. Production of oxygen, reducing carbon dioxide, maintaining the water cycle, protecting soil are important services. The world nowacknowledges that the loss of biodiversity contributes to global climatic changes. Forests arethe main mechanism for the conversion of carbondioxide into carbon and oxygen. The lossof forest cover, coupled with the increasing release of carbon dioxide and other gases through industrialization contributes to the **'greenhouseeffect'.** Global warming is melting ice caps, resultingn a rise in the sea level which will submergethe low lying areas in the world. It iscausing major atmospheric changes, leading to increased temperatures, serious droughts insome areas and unexpected floods in other areas. Biological diversity is also essential for preserving ecological processes, such as fixing and recycling of nutrients, soil formation, circulationand cleansing of air and water, global life support (plants absorb CO_2, give out O_2), maintaining the water balance within ecosystems, watershed protection, maintaining stream and river flows throughout the year, erosion control and local flood reduction.

Humans derive many direct and indirect benefits from the living world. Biodiversity is the source of food, medicines, pharmaceutical drugs, fibres, rubber and timber. Thebiological resources contain potentially useful resources as well. The diversity of organisms also provides many ecological services free of charge that are responsible for maintaining ecosystem health. The uses of biodiversity are briefly described below.

Consumptive use value

The biodiversity held in the ecosystem provides forest dwellers with all their daily needs, food, building material, fodder, medicines and a varietyof other products. They know the qualities and different uses of wood from different species of trees, and collect a large number of local fruits, roots and plant material that they use as food, construction material or medicines. Fisherfolk are highly dependent on fish and know where and how to catch fish and other edible aquatic animals and plants.

Social values

While traditional societies which had a small population and required less resources had preservedtheir biodiversity as a life supporting resource, modern man has rapidly depleted it evento the extent of leading to the irrecoverable lossdue to extinction of several species. Thus apart from the local use or sale of products of biodiversity there is the social aspect in whichmore and more resources are used by affluent societies. The biodiversity has to a great extent been preserved by traditional societies that valuedit as a resource and appreciated that its depletion would be a great loss to their society. The consumptive and productive value of biodiversity is closely linked to social concernsin traditional communities. ‘Ecosystem people’ value biodiversity as a part of their livelihood as well as through cultural and religious sentiments. A great variety of crops have been cultivated in traditional agricultural systems and this permitted a wide range of produce to be grown and marketed throughout the year and acted as an insurance against the failure of one crop. In recentyears farmers have begun to receive economic incentives to grow cash crops for national or international markets, rather than to supply local needs.

This has resulted in local food shortages, unemployment (cash crops are usually mechanised), landlessness and increased vulnerability to drought and floods.

Source of Food and Improved Varieties

Biodiversity is of use to modern agriculture in three ways: (i) as a source of new crops, (ii) as a source material for breeding improved varieties, and (iii) as a source of new biodegradable pesticides.

Of the several thousand species of edible plants, less than 20 plant species are cultivated to produce about 85% of the world's food. Wheat, corn and rice, the three major carbohydrate crops, yield nearly two-third of the food sustaining the human population. Fats, oils, fibres, etc. are other uses for which more and more new species need to be investigated.

The commercial, domesticated species are cross bred with their wild relatives to improve their traits. Genes of wild species are used to confer new properties such as disease resistance or improved yield in domesticated species. For example, rice grown in Asia is protected from the four main diseases by genes received from a single wild rice species (*Oryza nivara*) from India.

Drugs and Medicines

Biodiversity is a rich source of substances with therapeutic properties. Several important pharmaceuticals have originated as plant-based substances. Examples of plant-derived substances developed into valuable drugs are: Morphine (*Papaver somniferum*), used as an analgesic; Quinine (*Chinchona ledgeriana*) used for the treatment of malaria; and Taxol, an anticancer drug obtained from the bark of the yew tree (*Taxus brevifolia, T. baccata*). Currently, 25% of the drugs in the Pharmacy are derived from a mere 120 species of plants. But, throughout the world, traditional medicines make use of thousandsof plant species. Plants can also be used for the manufacture of innumerable synthetic products, called botanic chemicals.

Aesthetic and Cultural Benefits

Biodiversity has also great aesthetic value. Examples of aesthetic rewards include ecotourism, bird watching, wildlife, pet keeping, gardening, etc. Throughout human history, people have related biodiversity to the very existence of human race through cultural and religious beliefs. In a majority of Indian villages and towns, plants like *Ocimum sanctum* (Tulsi), *Ficus religiosa* (Pipal), and *Prosopis cineraria* (Khejri) and various other trees are planted, which are considered sacred and worshipped by the people. Several birds, and even snake, have been considered sacred. Today, we continue to recognize plants and animals as symbols of national pride and cultural heritage.

Megacenters of biodiversity and hot spots

Biodiversity is distributed heterogeneously across the Earth. Some areas are full with biological variations (e.g. tropical forests) others are virtually devoid of life (e.g. some deserts and polar regions) and most fall some where in between. The regions where alarge number of species are found are described as mega centres of biodiversity or mega diversity zone. India is recognized as one of the World's 12 mega diversity zones. India has over 45,000 species of flora and 75,000 species of fauna. India contributes nearly 8 % species to the global biodiversity in spite of having only 2.4 % of the land area of the world.

India's biodiversity

India is exceptionally rich in biodiversity and is one of the twelve mega diversity centres of the world. With 10 biogeographic zones and 25 biotic provinces, all major ecosystems are represented. India is a land mass of nearly 33 lakh sq.km with a coastline of 7,616 km and 14 different types of climatic forests and the total forest coverage in India is about 6,50,000 sq.km. India is the home land of 13,000 species of flowering plants, 20,000 species of fungi, 50,000 species of insects, 65,000 species of fauna including 2000 species of birds, 350 mammals and 420 of reptiles. It covers nearly 7% of world's flora and 6.5% of world's fauna of which 33 % flora and 62% fauna are endemic. India has over 30 National parks that constitute about

1% of the landmass and 441 sanctuaries that constitute 3.5% of the area. India is a home of over 35,000 tigers and the umbrella of project tiger 23 specially demarcated project tiger reserves covering 33,000 sq.km representing different climatic forests are spread across the country.

India has a rich and varied heritage of biodiversity, encompassing a wide spectrum of habitats from tropical rainforests to alpine vegetation and from temperate forests to coastal wetlands. India figured with two hotspots - the Western Ghats and the Eastern Himalayas - in an identification of 18 biodiversity hotspots carried out in the eighties. Recently, Norman Myers and a team of scientists have brought out an updated list of 25 hotspots. In the revised classification, the 2 hotspots that extend into India are The Western Ghats/Sri Lanka and the Indo-Burma region (covering the Eastern Himalayas); and they are included amongst the top eight most important hotspots. In addition, India has 26 recognised endemic centres that are home to nearly a third of all the flowering plants identified and described to date.

Of the estimated 5–50 million species of the world's biota, only 1.7 million have been described to date, and the distribution is highly uneven.

About seven per cent of the world's total land area is home to half of the world's species, with the tropics alone accounting for 5 million. India contributes significantly to this latitudinal biodiversity trend. With a mere 2.4% of the world's area, India accounts for 7.31% of the global faunal total with a faunal species count of 89,451 species. Some salient features of India's biodiversity have been mentioned below.

- India has two major realms called the Palaearctic and the Indo-Malayan, and three biomass, namely the tropical humid forests, the tropical dry/deciduous forests, and the warm desert/semi-deserts
- India has ten biogeographic regions including the Trans-Himalayan, the Himalayan, the Indian desert, the semi-arid zone(s), the Western Ghats, the Deccan Peninsula, the Gangetic Plain, North-East India, and the islands and coasts.

- As of date, there are 911 properties under the World Heritage List, which cover 711 cultural sites, 180 natural sites and 27 mixed properties encompassing 152 countries, including India. India is one of the 12 centres of origin of cultivated plants.
- India's first two sites inscribed on the list at the Seventh Session of the World Heritage held in 1983 were the Agra Fort and the Ajanta Caves. Over the years, 27 more sites have been inscribed, the latest site inscribed in 2012 being the Western Ghats. Of these 29 sites, 23 are cultural sites and the other six are natural sites. A tentative list of further sites/properties submitted by India for recognition includes 33 sites.
- India has 17 biosphere reserves, and 19 Ramsar wetlands. Amongst the protected areas, India has 102 national parks and 490 sanctuaries covering an area of 1.53 lakh sq. km.
- The wildlife sanctuaries in India are home to around two thousand different species of birds, 3500 species of mammals, nearly 30000 different kinds of insects and more than 15000 varieties of plants

The endemism of Indian biodiversity is high. About 33% of the country's recorded flora are endemic to the country and are concentrated mainly in the North-East, Western Ghats, North-West Himalaya and the Andaman and Nicobar islands. Of the 49,219 plant species, 5150 are endemic and distributed into 141 genera under 47 families corresponding to about 30% of the world's recorded flora, which means 30% of the world's recorded flora are endemic to India. Of these endemic species, 3,500 are found in the Himalayas and adjoining regions and 1600 in the Western Ghats alone. About 62% of the known amphibian species are endemic with the majority occurring in the Western Ghats.

Nearly 50% of the lizards of India are endemic with a high degree of endemicity in the Western Ghats. India is a centre of crop diversity - the homeland of 167 cultivated species and 320 wild relatives of crop plants.Corals reefs in Indian waters surround the Andaman and Nicobar Islands, the Lakshadweep Islands, and the Gulf areas of Gujarat and Tamil Nadu. They are nearly as rich in species as tropical evergreen forests. India's record in agro-biodiversity is equally impressive. There are 167 crop species and wild relatives. India is considered to be the centre of origin of 30,000-50,000 varieties of rice, pigeon-pea, mango, turmeric, ginger, sugarcane, gooseberries etc and ranks seventh in terms of contribution to world agriculture.

Threats to biodiversity

Important factors leading to extinction of species and consequent loss of biodiversity are: habitat loss and fragmentation, introduction of non-native species, overexploitation, soil,water and atmospheric pollution, and intensive agriculture and forestry.

a) Habitat Loss and Fragmentation

The destruction of habitats is the primary reason for the loss of biodiversity. When peoplecut down trees, fill a wetland, plough a grassl and or burn a forest, the natural habitat of a species is changed or destroyed. These changes can kill or force out many plants, animals, and microorganisms, as well as disrupt complex interactions among the species.A forest patch surrounded by croplands, orchards, plantations, or urban areas is anexample of fragmented habitats. With the fragmentation of a large forest tract, species occupying deeper parts of forests are the first to disappear. Over exploitation of a particular species reduces the size of its population to an extent that it becomes vulnerable to extinction.

b) Disturbance and Pollution

Communities are affected by natural disturbances, such as fire, tree fall, and defoliation by insects. Man-made disturbances differ from natural disturbances in intensity, rate and spatial extent. For example, man by using fire more frequently may change speciesrichness of a community. Then, some human impacts are new, never before faced by biota, e.g. the vast number of synthetic compounds, massive releases of radiation or spillover of oil in sea. These impacts lead to a change in the habitat quality. Pollution may reduce and eliminate populations of sensitive species. For example, pesticide linked decline of fish-eating birds and falcons. Lead poisoning is another major cause of mortality of many species, such as ducks, swans and cranes, as they ingest the spent shotgun pellets that fall into lakes and marshes.

Eutrophication (nutrient enrichment) of water bodies drastically reduces species diversity.

c) Introduction of Exotic Species

New species entering a geographical region are called exotic or alien species. Introduction of such invasive species may cause disappearance of native species through changed biotic interactions. Invasive species are considered second only to habitatdestruction as a major cause of extinction of species. Exotic species are having large impact especially in island ecosystems, which harbour much of the world's threatened biodiversity. A few examples are:

(1) Nile perch, an exotic predatory fish introduced into Lake Victoria (South Africa) threatens the entire ecosystem of the lake by eliminating several native species ofthe small Cichlid fish species that were endemic to this freshwater aquatic system.

(2) Water hyacinth clogs rivers and lakes and threatens the survival of many aquaticspecies in lakes and river flood plains in several tropical countries including India.

(3) *Lantana camara* has invaded many forest lands in different parts of India, and strongly competes with the native species.

d) Extinction of Species

Extinction is a natural process. Species have disappeared and new ones have evolved to take their place over the long geological history of the earth. It is useful to distinguish three types of extinction processes.

i) **Natural extinction:** With the change in environmental conditions, some species disappear and others, which are more adapted to changed conditions, take their place.This loss of species which occurred in the geological past at a very slow rate is called natural or background extinction.

ii) **Mass extinction:** There have been several periods in the earth's geological history when large number of species became extinct because of catastrophes. Mass extinctions occurred in millions of years.

iii) **Anthropogenic extinction:** An increasing number of species is disappearing from the face of the earth due to human activities. This man-made mass extinction represents avery severe depletion of biodiversity, particularly because it is occurring within a short period of time.

Conservation of biodiversity

We know that ecosystems are undergoing change due to pollution, invasive species, overex ploitation by humans, and climate change. Most people are beginning to recognize that diversity at all levels - gene pool, species and biotic community is important and needs to be conserved. We should not deprive the future generations from the economic and aesthetic benefits that they can derive from biodiversity. The decisions we make now, as individuals and as a society, will determine the diversity of genes, species and ecosystems that remain in future. We may appreciate the fact that the most effective andefficient mechanism for conserving biodiversity is to prevent further destruction or degradation of habitats by us. We require more knowledge to conserve biodiversity inreduced space and under increased pressure of human activities. There are two basic strategies of biodiversity conservation, *in situ* (on site) and *ex situ* (off site).

A. *In situ* Conservation Strategies

The *in situ* strategy emphasizes protection of total ecosystems. The *in situ* approach includes protection of a group of typical ecosystems through a network of protected areas.

Protected areas: These are areas of land and/or sea especially dedicated to the protection and maintenance of biological diversity, and of natural and associated cultural resources. These are managed through legal or other effective means. Examples of protected areasare National Parks, and Wildlife Sanctuaries. World Conservation Monitoring Centre has recognized 37000 protected areas around the world. As of September 2002, India has 581 protected areas (89 National Parks and 492 Wildlife Sanctuaries), covering 4.7% of the land surface as against 10% internationally suggested norm. The Jim Corbett National Park was the first National Park established in India.

Some of the main benefits of protected areas are:

(1) Maintaining viable populations of all native species and subspecies;
(2) Maintaining the number and distribution of communities and habitats, and conserving the genetic diversity of all the present species;
(3) Preventing human-caused introductions of alien species; and
(4) Making it possible for species/habitats to shift in response to environmental changes.

Biosphere Reserves: Biosphere reserves are a special category of protected areas of land and/or coastal environments, wherein people are an integral component of the system. These are representative examples of natural biomes and contain unique biological communities. The concept of Biosphere Reserves was launched in 1975 as a part of the UNESCO's Man and Biosphere Programme dealing with the conservation of ecosystems and the genetic resources contained therein. Till May 2002, there were 408 biosphere reserves located in 94 countries. India now has 14 biosphere reserves, 90 national parks and 448 wildlife sanctuaries. In India, Biosphere Reserves are also notified as National Parks. India has also a history of religious and cultural traditions that emphasised protection of nature.

A Biosphere Reserve consists of core, buffer and transition zones. The natural or corezone comprises an undisturbed and legally protected ecosystem. The buffer zone surrounds the core area, and is managed to accommodate a greater variety of resource usestrategies, and research and educational activities. The transition zone, the outermost part of the Biosphere Reserve, is an area of active cooperation between reserve management and the local people, wherein activities like settlements, cropping, forestry and recreation and other economic uses continue in harmony with conservation goals.

The main functions of biosphere reserves are:

(1) Conservation: to ensure the conservation of landscapes, ecosystems, species and genetic resources. It also encourages traditional resource use.
(2) Development: to promote economic development, which is culturally, socially and ecologically sustainable.
(3) Scientific research, monitoring and education: the aim is to provide support for research, monitoring, education and information exchange related to local, national and global issues of conservation and development.

Sacred forests and sacred lakes: A traditional strategy for the protection of biodiversity has been in practice in India and some other Asian countries in the form of sacred forests.These are forest patches of varying dimensions protected by tribal communities due to religious sanctity accorded to these forest patches. The sacred forests represent islands of pristine forests (most undisturbed forests without any human impact) and have been freefrom all disturbances; though these are frequently surrounded by highly degraded landscapes. In India sacred forests are located in several parts, e.g. Karnataka, Maharashtra, Kerala, Meghalaya, etc., and are serving as refugia for a number of rare,endangered and endemic taxa. Similarly, several water bodies (e.g. Khecheopalri Lake in Sikkim) have been declared sacred by the people leading to protection of aquatic flora and fauna.

B. *Ex-situ* Conservation Strategies

The *ex-situ* conservation strategies include botanical gardens, zoos, conservation stands, and gene, pollen, seed, seedling, tissue culture and DNA banks. Seed gene banks are the easiest way to store germplasm of wild and cultivated plants at low temperature in cold rooms. Preservation of genetic resources is carried out in field gene banks under normalgrowing conditions.

In vitro conservation, especially by cryopreservation in liquid nitrogen at a temperature of –196°C, is particularly useful for conserving vegetatively propagated crops like potato. Cryopreservation is the storage of material at ultra-low temperature either by very rapid cooling (used for storing seeds) or by gradual cooling and simultaneous dehydration at low temperature (used for tissue culture). The material can be stored for a long period of time in compact, low maintenance refrigeration units.

The conservation of wild relatives of crop plants and the off-site conservation of crop varieties or cultures of microorganisms provide breeders and genetic engineers with a ready source of genetic material. Plants and animals conserved in botanical gardens, arboreta, zoos and aquaria can be used to restore degraded land, reintroduce species intowild, and restock depleted populations.

Biodiversity conservation in India

Indian region has contributed significantly to the global biodiversity. India is a homeland of 167 cultivated species and 320 wild relatives of crop plants. It is the centre of diversity of animal species (zebu, mithun, chicken, water buffalo, camel); crop plants(rice, sugarcane, banana, tea, millet); fruit plants and vegetables (mango, jackfruit,cucurbits), edible diascoreas, alocasia, colocasia; spices and condiments (cardamom,black pepper, ginger, turmeric); bamboos, brassicas, and tree cotton. India also represents a secondary centre of domestication for some animals (horse, goat, sheep, cattle, yak, and donkey) and plants (tobacco, potato and maize).

The National Bureau of Plant, Animal and Fish Genetic Resources has a number of programmes to collect and conserve the germplasm of plants and animals in seed genebanks and field gene banks for *in vitro* conservation. Botanical and zoological gardens have large collections of plant and animal species in different climatic regions of India. The land races and diverse food and medicinal plants are also being conserved successfully by the tribal people and women working individually or with various non governmental agencies. The women particularly have an important role in the conservation of agrobiodiversity. In India, a programme is underway to develop a system of community registers of local informal innovations related to the genetic resources as well as natural resource management in general.

Biological Diversity act 2002

Biological diversity is a national asset of a country; hence the conservation of biodiversity assumes greater significance. The first attempt to bring the biodiversity into the legal frame work was made by way of the biodiversity bill 2000 which was passed by the Lok sabha on 2nd December 2002 and by Rajya Sabha on the December 2002.

Objectives of the act

1. To conserve the Biological Diversity
2. Sustainable use of the components of biodiversity
3. Fair and equitable sharing of benefits arising out of the use of the B.D

A national biodiversity authority has been established by the Biodiversity Act 2002 to regulate act implementing rules 2004 has been operationalised since coming into force.

Act: Regulating access well as pushing the officially sponsored documentation of biological resources and traditional practices through people's diversity registers at the local and data bases at the national levels, respectively. It further probes the extent to which the principles of conservation have realized.

Provisions of Act.

1. Prohibition on transfer of Indian genetic material outside the country without specific
   a. approval of the Indian Government
2. Prohibition of anyone claiming an (IPR) such as a patent over biodiversity or related knowledge without permission of Indian Government.
3. Regulation of collection and use of biodiversity by Indian national while exemptinglocal communities from such restrictions
4. Measures from sharing of benefits from use of biodiversity including transfer of technology, monetory returns, joint research and development, joint IPR ownership etc.

5. Measuring to conserve sustainable use of biological resources including habitat and species protection (EIP) of projects, integration of biodiversity into the plans and policies of various Departments and Sectors.
6. Provisions for local communities to have a say in the use of their resources and knowledge and to charge fees for this
7. Protection of indigenous or tradition laws such as registration of such knowledge
8. Regulation of the use of the genetically modified organisms
9. Setting up of National, state and local Biodiversity funds to be used to support conservation and benefit sharing
10. Setting up of Biodiversity Management committees (BMC) at local village levels, State Biodiversity Boards at state level and National Biodive rsity Authority.

Functions of Authority

1. Advise the central Government on any matter concerning conservation of biodiversity sustainable use of its components and fair and equitable sharing of benefits arising out of the use of biological resource and knowledge
2. Coordinate the activities of state biodiversity
3. Provide the technical assistance and guidance to the state biodiversity boards
4. Sponsor investigation and research
5. Engage consultants for a specific period not exceeding 3 years for providing technical assistance to the Authority in the effective discharges of its functions.
6. Collect, compile and publish technical and statistical data, manuals, codes or guides relating to conservation of biodiversity, sustainable use of its components and fair and equitable sharing of benefits arising out of the use of biological resource and knowledge
7. Organize through mass media a comprehensive programme regarding conservation of biodiversity, sustainable use of components and fair and equitable sharing of benefits arising out of the use of biological resources and knowledge.

8. Plan and organize training of personal engaged or likely to be engaged in programmes for the conservation of biodiversity and sustainable use of its components
9. Prepare the annual budget of the authority including its own receipts as also the devaluation from the central Government provided that the allocation by the central government shall be operated in accordance with budget provisions approved by the central govt.
10. Recommend creation of posts to the central Government for effective discharge of the functions by the authority.
11. Approve the method of recruitment to the officers and servants of the authority.
12. Take steps to build up database and to create information and documentation system for biological resources and associated traditional knowledge through biodiversity register and electronic databases to ensure effective management, promotion and sustainable uses.
13. Give directions to state Biodiversity Boards and the Biodiversity Management Committees in writing for effective implementation of the act.
14. Report to the central Government about the functioning of the Authority and implementation of the Act
15. Sanction grants to the State Biodiversity Board and Biodiversity Management committees for specific purposes.
16. Take necessary measures including appointment of legal experts to oppose grant of intellectual property right in any country outside India on any biological outside India on any biological resource and associated knowledge obtained from India and in an illegal manner.
17. Do such other functions as may be assigned to directed by the central government from time to time
18. Regulates the commercial utilization or biosurvey and bio-utilization of any biological resource by Indians.

Agrobiodiversity

Agrobiodiversity is the result of natural selection processes and the careful selection and inventive developments of farmers, herders and fishers over millennia. Agrobiodiversity is a vital sub-set of biodiversity. Many people's food and livelihood security depend on the sustained management of various biological resources that are important for food and agriculture.

Indian gene centre is among the 12 mega diversity regions of the world and about25 crop species were domesticated here. It is known to have more than 18,000 species of higher plants including, 160 major and minor crop species and 325 of their wild relatives.In addition, ethnobotanical uses of nearly 9,500 plant species have been reported from the country, of which around 7,500 are for ethno medicinal purposes and 3,900 are multipurpose/edible species. Around 1,500 wild edible plant species are widely exploited by native traditional people including 145 species of roots and tubers, 521 of leafyvegetables/greens, 101 of buds and flowers, 647 of fruits and 118 of seeds and nuts.

Agricultural biodiversity, also known as agrobiodiversity or the genetic resources for food and agriculture, includes:

- Harvested crop varieties, livestock breeds, fish species and non domesticated (wild) resources within field,forest, rangeland including tree products, wild animals hunted for food and in aquatic ecosystems (e.g. wild fish);
- Non-harvested species in production ecosystems that support food provision, including soil micro-biota, pollinators and other insects such as bees, butterflies, earthworms, greenflies; and
- Non-harvested species in the wider environment that support food production ecosystems (agricultural, pastoral, forest and aquatic ecosystems).

Agrobiodiversity is the result of the interaction between the environment, genetic resources and management systems and practices used by culturally diverse peoples, and therefore land and water resources are used for production in different ways. Thus, agrobiodiversity encompasses the variety and variability of animals, plants and micro-organisms that are necessary for sustaining key functions of the agro-ecosystem, including its structure and processes for, and in support of, food production and food security. Local knowledge and culture can therefore be considered as integral parts of agrobiodiversity, because it is the human activity of agriculture that shapes and conserves this biodiversity. Here organic agriculture plays a major role in maintaining the biodiversity of agriculture field.

There are several distinctive features of agrobiodiversity, compared to other components of biodiversity:

- Agrobiodiversity is actively managed by male and female farmers;
- Many components of agrobiodiversity would not survive without this human interference; local knowledge and culture are integral parts of agrobiodiversity management;
- Many economically important agricultural systems are based on 'alien' crop or livestock species introduced from elsewhere (for example, horticultural production systems or Friesian cows in Africa). This creates a high degree of interdependence between countries for the genetic resources on which our food systems are based;
- As regards crop diversity, diversity within species is at least as important as diversity between species;
- Because of the degree of human management, conservation of agrobiodiversity in production systems is inherently linked to sustainable use – preservation through establishing protected areas is less relevant; and
- In industrial-type agricultural systems, much crop diversity is now held *ex situ* in gene banks or breeders' materials rather than on-farm.

Biological diversity (or biodiversity) can be measured on three levels: genetic diversity, species diversity and ecosystem diversity. Agricultural biodiversity, or agro-biodiversity, is a vital sub-set of biodiversity.

The interactions between farming and biodiversity are complex. At the regional level, the type of agriculture (i.e. intensive arable crops, dairy farming) and the area which is cultivated (i.e. mountain pastures, grasslands) are the main determinants. At a landscape level, patterns of field-size, land cover and types of field boundaries all effect biodiversity. At the farm level, land use management (e.g. crop succession and rotations) is a key factor. At the field level, different farming practices determine habitat quality and associated biodiversity. At each level, different ecological processes affect species distribution.

What is happening to agrobiodiversity?

Locally varied food production systems are under threat, including local knowledge and the culture and skills of women and men farmers. With this decline, agrobiodiversity is disappearing; the scale of the loss is extensive. With the disappearance of harvested species, varieties and breeds, a wide range of unharvested species also disappear.

100 years of agricultural change:Trends and figures related to agrobiodiversity

- Since the 1900s, some 75 percent of plant genetic diversity has been lost as farmers worldwide have left their multiple local varieties and landraces for genetically uniform, high-yielding varieties.
- 30 percent of livestock breeds are at risk of extinction; six breeds are lost each month.
- Today, 75 percent of the world's food is generated from only 12 plants and five animal species.
- Of the 4 percent of the 250 000 to 300 000 known edible plant species, only 150 to 200 are used by humans. Only three – rice, maize and wheat – contribute nearly 60 per cent of calories and proteins obtained by humans from plants.
- Animals provide some 30 percent of human requirements for food and agriculture and 12 per cent of the world's population live almost entirely on products from ruminants.

More than 90 per cent of crop varieties have disappeared from farmers' fields; half of the breeds of many domestic animals have been lost. In fisheries, all the world's 17 main fishing grounds are now being fished at or above their sustainable limits, with many fish populations effectively becoming extinct. Loss of forest cover, coastal wetlands, other 'wild' uncultivated areas, and the destruction of the aquatic environment exacerbate the genetic erosion of agrobiodiversity.

Fallow fields and wildlands can support large numbers of species useful to farmers. In addition to supplying calories and protein, wild foods supply vitamins and other essential micronutrients. In general, poor households relyon access to wild foods more than the wealthier. However, in some areas, pressure on the land is so great that wild food supplies have been exhausted.

The term 'wild-food', though commonly used, is misleading because it implies the absence of human influence and management. Over time, people have indirectly shaped many plants. Some have been domesticated in homegardens and in the fields together with farmers' cultivated food and cash crops. The term 'wild-food', therefore, is used to describe all plant resources that are harvested or collected for human consumption outside agricultural areas in forests, savannah and other bush land areas. Wild-foods are incorporated into the normal livelihood strategies of many rural people, pastoralists, shifting cultivators, continuous croppers or hunter-gatherers. Wild-food is usually considered as a dietary supplement to farmers' daily food consumption, generally based on their crop harvest,domestic livestock products and food purchases on local markets. For instance, fruits and millets, from a wide range of wild growing plants, are typically referred to as 'wild-food'. Moreover, , fruits and millets add crucial vitamins tothe normally vitamin deficient country cereal diet, particularly for children.

There are many reasons for this decline in agrobiodiversity. Throughout the twentieth century the decline has accelerated, along with increased demands from a growing population and greater competition for natural resources.

The principal underlying causes include:

The rapid expansion of industrial and Green Revolution agriculture: This includes intensive livestock production, industrial fisheries and aquaculture. Some production systems use genetically modified varieties and breeds. Moreover, relatively few crop varieties are cultivated in monocultures and a limited number of domestic animal breeds, or fish, are reared or few aquatic species cultivated.

Globalization of the food system and marketing. The extension of industrial patenting, and other intellectual property systems, to living organisms has led to the widespread cultivation and rearing of fewer varieties and breeds. This results in a more uniform, less diverse, but more competitive global market. As a consequence there have been:

- Changes in farmers' and consumers' perceptions, preferences and living conditions;
- Marginalization of small-scale, diverse food production systems that conserve farmers' varieties of crops and breeds of domestic animals;
- Reduced integration of livestock in arable production, which reduces the diversity of uses for which livestock are needed; and,
- Reduced use of 'nurture' fisheries techniques that conserve and develop aquatic biodiversity.

The main cause of the genetic erosion of crops – as reported by almost all countries – is **the replacement of local varieties by improved or exotic varieties and species**. Frequently, genetic erosion occurs as old varieties in farmers' fields are replaced by newer. Genes and gene complexes, found in the many farmers' varieties, are not contained in the modern. Often, the number of varieties is reduced when commercial varieties are introduced into traditional farming systems. While FAO (1996) states that some indicators of genetic erosion have been developed, few systematic studies of the genetic erosion of crop genetic diversity have been made. Furthermore, in the FAO Country Reports (1996) nearly all countries confirm genetic erosion is taking place and that it is a serious problem.

Organic agriculture and biodiversity

Organic farming is a system of agriculture that relies largely on locally available resources and is dependent upon maintaining ecological balances and developing biological processes to their optimum. These systems take local soil fertility as a key to successful production. Wild species perform a variety of ecological services within organic agriculture: for example pollinators, natural enemies of pests and soil microorganisms are all key components in agro-ecosystems. Thus, higher levels of biodiversity can strengthen some farming systems and practices. Organic systems dramatically reduce external inputs by refraining from the use of synthetic chemical fertilisers, pesticides and pharmaceuticals. Instead, systems are designed to manage nature in order to determine agricultural yields and disease resistance. By respecting the natural capacity of plants, animals and the landscape, organic agriculture aims to optimize quality in all aspects of agriculture and the environment.

Organic agriculture is thus committed to the conservation of biodiversity within agricultural systems, both from a philosophical perspective and from the practical viewpoint of maintaining productivity. Biological pest control on organic farms, for example, relies on maintaining healthy populations of pest predators. By using a system of crop rotation, in time (over several years rotations) or in space (through intercropping or growing several crops in the same season in different fields), the build-up of harmful pests and diseases can be reduced and biodiversity increased, although with respect to grassland ecosystems it must be stressed that periodic ploughing is harmful to grassland vegetation, a reason grassland areas should, if possible, be kept as permanent grassland. One of the most important elements in conversion to organic systems has proved to be the time needed to restore a natural ecological balance with respect to pest-predator populations. Compared to conventional farms, 2-3 times more individual birds, greater number of earthworms and biomass; more individuals species of spiders; more non-pest species of butterflies were found on organic farms. The diversity (population or species or individuals) of vascular plants, different invertebrate groups and birds was 0.5–20 times higher on organic than conventional farms.

Organic agriculture also encourages variety. Reduced reliance on agrochemicals to control changes in soil conditions means that the plants must themselves be better adapted to local conditions. Organic systems thus encourage the expansion of varieties grown, and the preservation of older, locally bred varieties and breeds.

Research into organic agriculture practices has shown that organic systems have the potential to support biodiversity conservation through:

- Increasing the number and variety of wild species found on farms - Providing food and shelter for wild species found on farms and thus increasing them in number and variety
- Supporting high levels of agro-biodiversity
- Maintaining healthy soils and soil fauna, such as earthworms
- Reducing the risk of water pollution
- Being energy efficient
- Lowering emissions of carbon dioxide to reduce global warming
- About a third of the world's land surface is used for agriculture.
- Organic agriculture standards and practices ensure this area is sympathetically managed for biodiversity and that primary ecosystems are not cleared to further extend the agricultural frontier.

Conservation of cultivars and livestock breeds

Fifty years ago nearly 30,000 rice var. were grown in India now only a few of these are cultivated. The new varieties being developed use the germplasm of these original types. But if all these traditional types vanish, it would be difficult to develop new disease resistant varieties for future. Use of varieties from gene banks have been expensive and risky. Farmers need to be encouraged to grow traditional varieties. This is a concern for future of mankind. Gene banks have at present 34,000 creeds and 2200 pulses). Traditional breeds/ varieties have to be encouraged for genetic variability. Incontrast men interested in cash returns in short time wouldn't appreciate the benefits ofgrowing indigenous varieties.Hardy cultivars and breeds are bases of organic cropping systems and thus contribute to the preservation of locally-adapted (and therefore diversified) breeds and cultivars; population varieties and plant mixtures are also used. This also presents an opportunity to develop participative plant breeding towards ideotypes or ideotype assemblages besideusual standards

Enhances Pollinator Populations

Research found that native bee populations supported 50-100% of the pollination needs for a watermelon crop on organic farms and none of the pollination needs on non-organic farms, which required supplemental pollination from honey bees. The study also noted that the proximity of a field to natural habitat was a factor in influencing native pollinator services, regardless of whether the field was organically or non-organically managed. A Canadian study showed that organic canola fields were found to have greater abundance of native bee communities than non-organic fields.

Enhances Bird and Beneficial Spider Populations

A two-year study in Nebraska found that fields on organic farms had both more birds and more bird species than were found on non-organic farms, while Florida research found that the practice of intercropping sunflower into organic vegetable fields increased "incidence, abundance, and foraging activity" by insect-eating native birds. A study of apple orchards in Washington comparing synthetic, broad-spectrum pesticides with organic management showed the total arboreal and understory spider populations were significantly higher in the organic orchards. They conclude "spider populations may be severely reduced by even a small number of synthetic, broad-spectrum insecticide applications and the time required for recovery may be lengthy.

Enhances Natural Enemy Populations

A study from California reveals that there was higher natural enemy abundance and greater species richness of all groups of insects on organic than non- organic farms, meaning that crop pests would encounter more potential predators on organic than on non-organic farms

Enhances soil biodiversity

Organic fertilising inputs and the use of compost supply the soil with organic matter which increases soil organic content, thus favouring detritivore organisms and the permanency of food webs. Besides plant nutrition, such processes are highly favourableto biodiversity and functional biodiversity. Tillage practices also contribute tosoil aeration and are favourable to below ground living organisms.

- **Abundant arthropods and earthworms.** Organic management increases the abundance and species richness of beneficial arthropods living above ground and earthworms, and thus improves the growth conditions of crops. More abundant predators help to control harmful organisms (pests). In organic systems the density and abundance of arthropods, as compared to conventional systems, has up to 70-120% more spiders. The biomass of earthworms in organic systems is 30-40% higher than in conventional systems, their density even 50-80% higher. Compared to the mineral fertilizer system, this difference is even more pronounced.
- **High occurrence of symbionts.** Organic crops profit from root symbioses and are better able to exploit the soil. On average, mycorrhizal colonization of roots is highest in crops of unfertilized systems, followed by organic systems. Conventional crops have colonization levels that are 30% lower. Even when all soils are inoculated with active mycorrhizae, colonization is enhanced in organic soil.
- **High occurrence of micro-organisms.** Earthworms work hand in hand with fungi, bacteria, and numerous other microorganisms in soil. In organically managed soils, the activity of these organisms is higher. Microorganisms in organic soils not only mineralize more actively, but also contribute to the build up of stable soil organic matter (there is less untouched straw material in organic than in conventional soils). Thus, nutrients are recycled faster and soil structure is improved. The amount of microbial biomass and decomposition is connected: at high microbial biomass levels, little light fraction material remains undecomposed and vices versa.
- **Microbial carbon.** The total mass of microorganisms in organic systems is 20-40% higher than in the conventional system with manure and 60-85% than in the conventional

system without manure. The ratio of microbial carbon to total soil organic carbon is higher in organic system as compared to conventional systems. Organic management promotes microbial carbon (and thus, soil carbon sequestration potential).

- **Enzymes.** Microbes have activities with important functions in the soil system: soil enzymes indicate these functions. The total activity of microorganisms can be estimated by measuring the activity of a living cell-associated enzyme such as dehydrogenase. This enzyme plays a major role in the respiratory pathway. Proteases in soil, where most organic N is protein, cleave protein compounds. Phosphatases cleave organic phosphorus compounds and thus provide a link between the plant and the stock of organic phosphorus in the soil. Enzyme activity in organic soils is markedly higher than in conventional soils. Microbial biomass and enzyme activities are closely related to soil acidity and soil organic matter content.
- **Wild flora.** Large organic fields (over 15 ha) featured flora six times more abundant than conventional fields, including endangered varieties. In organic grassland, the average number of herb species was found to be 25 percent more than in conventional grassland, including some species in decline.

Chapter 5

Good Agricultural Practices (GAP)

The increasing awareness of the deleterious effects of indiscriminate use of artificial inputs in agriculture such as chemical fertilizers and pesticides hasled to the adoption of organic farming as an alternative method for conventional farming. Farmers are adopting organic farming due to the advantages like,

- It is self-sustaining and socially and ecologically superior over the conventional farming.
- It makes use of cost effective management practices involving the use of farm inputs produced within the farm.
- It is less expensive.
- It is environment friendly as there is no pollution of soil and water.
- It enriches the soil and the local ecology.
- The produce is free of contamination from chemical residues has better taste, flavour and nutritional value.
- Seeds have more vitality and are good for successive generations.
- Growing markets for safe food

Food safety has gained increasing importance over the years because of its impact on the health of consumers and the growth in the domestic and international trade in food products. Production of safe food is essential for protecting consumers from the hazards of foodborne illnesses. Furthermore, food safety is an integral part of food security and also contributes towards increasing competitiveness in export markets. Hazards in food may be introduced at different stages of the food chain starting right from the primary production, e.g. residues above permitted levels, microbial contaminants and heavy metals. It therefore becomes important to address food safety right from food production at farm level. Implementing Good Agricultural Practices (GAP) during on-farm production and post-production processes resulting in safe agricultural products is of immense importance for ensuring a safe food supply.

It is prerequisite to understand and follow good agricultural practices (GAP) before adopting organic agriculture. This would help the farmer to plan andprepare the needed inputs by efficiently utilizing the local resources. The concept of Good Agricultural Practices (GAP) has evolved in recent years in the context of a rapidly changing and globalizing food economy. In addition, the concerns and commitments of a wide range of stake holders on food production and security, food safety and quality and the environmental sustainability of agriculture have also promoted the concept of GAP. These stakeholders include governments, food processing and retailing industries, farmers and consumers. According to the Food and Agriculture Organization (FAO), GAP is the application of available knowledge to address environmental, economic andsocial sustainability for on-production and post-production process resulting in production of safe and healthy food and non-food agricultural products. Many farmers in developed and developing countries already apply GAP through sustainable agricultural methods such as integrated pest management, integrated nutrient management and conservation agriculture. These methods are applied in a range of farming systems and scales of production units, facilitated by supportive government policies.

GAP is one of the most important contributors to the preventative practices mentioned earlier and ensures that on-farm practices result in safe produce reaching the farm gate. GAP is a practice that needs to be applied on the farm to ensure food safety during the pre-production, production, harvest and post-harvest stages. In many cases, such practices also help to protect the environment and safety of the workers. In other words GAP is a systematic approach that aims at applying available knowledge to address environmental, economic and social sustainability dimensions of on-farm production and post-production processes, resulting in safe and quality food and non-food agricultural products.

Now GAP is formally recognized in the international regulatory framework for reducing risks associated with the use of pesticides, taking into account public and occupational health, environmental and safety considerations. This increasing trend of acceptance of GAP by the consumers and the retailers provides incentives to the farmers by paying a premium where in farmer would find alternatives to reduce the contamination right from the sowing of crop to harvest. In addition, implementing GAP also helps promote sustainable agriculture and contributes to meeting national and international environmental and social developmental objectives. GAP applies to a wide range of food/ agricultural commodities that include fruits and vegetables, dairy products, medicinal and aromatic herbs, ornamentals, aquaculture etc.

Risk Minimizing Measures following GAP

a) Pre-planting Measures

i. Site Selection

Land or site for agricultural production selected is on the basis of land history, previous manure applications and crop rotation.

ii. Manure handling and field application

Proper and thorough composting of manure, incorporating it into soil prior to planting and avoiding top dressing on plants are important steps to be followed.

iii. Manure storage and sourcing

Manure is stored in shade with sufficient aeration. It is important that duringthe aerobic composting process, high temperature to be achieved to kill most harmful pathogens.

iv. Timely application of manure

Manure should be applied at the end of the season to all planned vegetable ground or fruit orchard. If applied at the start of a season, then it should be spread two weeks before planting, preferably to grain or forage crops.

v. Selection of appropriate crops

A variety of crops adapted to the local area to be cultivated in an area. It is highly recommended to begin with a leguminous crop or vegetables instead ofcereals or high nutrient demanding crops.

b) Production Measures

i. Irrigation Water Quality

Irrigation water should be free from pathogens and pesticide residues.Surface water is tested quarterly in the laboratory for any contaminations.Farmers can filter or use the settling ponds to improve water quality.

ii. Irrigation Methods

It is always advisable to use drip irrigation to reduce the contamination because the edible parts of most crops are not wetted directly. It alsoenhances the water use efficiency.

Good practices related to water will include those that maximize water infiltration and minimize unproductive efflux of surface waters from watersheds; manage ground and soil water by proper use, or avoidance of drainage where required; improve soil structure and increase soil organic matter content; apply production inputs, including waste or recycled products of organic, inorganic and synthetic nature by practices that avoid contamination of water resources; adopt techniques to monitor crop and soil water status, accurately schedule irrigation, and prevent soil salinization by adopting water-saving measures and re-cycling where possible; enhance the functioning of the water cycle by establishing permanent cover, or maintaining or restoring wetlands as needed; manage water tables to prevent excessive extraction or accumulation; and provide adequate, safe, clean watering points for livestock.

iii. Field sanitation

Great care to be taken to prevent the spread of human and animal pathogens.Animals especially poultry are not allowed to roam in the field especially close to the harvest time.

iv. Worker facilities and hygiene

The farm workers are provided with hygienic and well maintained toilet facilities around the farming areas. Farmers should get proper training to make them understand the relationship between food safety and personal hygiene. These measures are to be monitored and enforced.

c) Harvest Measures

i. Clean harvest Aids

Baskets, bins and all crop containers have to be washed and rinsed properly. They should be properly covered when not in use to avoid contamination by birds and animals.

ii. Worker hygiene and training

Good personal hygiene is very important during the harvest of the crops. Employees' awareness, meaningful training and accessibility to restroom facilities with hand wash stations encourage good hygiene.

d) Post-harvest Measures

i. Worker hygiene

Packaging area should be clean and sanitized. The worker should be clean and use disposable gloves on packing lines.

ii. Monitor wash water quality

Potable water should be preferably used in all washing operations. Use chlorinated water to wash the fresh produce.

iii. Sanitize packing house and packing operations

Loading, staging and all food contact surfaces should be cleaned and sanitized at the end of each day. Care is taken to prevent rats and rodents from entering the packing house. Packaging material should be stored in a clean area.

iv. Pre-cooling and cold storage

Harvested produce should be quickly cooled to minimize the growth of pathogens and maintain good quality. Refrigeration room should not be overloaded beyond cooling capacity.

v. Transportation of produce from farm to produce

Cleanliness of the transportation vehicles is to be maintained. For traceability norms, it must be ensured that each package leaving the farm can be traced to field of origin and date of packing.

Chapter 6

Organic Sources of Plant Nutrients & Its Management

Soil is a living system and soil fertility is the key to agricultural productivity. Any input that destroys this living system and undermines soil health basically undermines the agricultural productivity. The maintenance of the fertility of soil is the primary step in any permanent system of agriculture.

Fertile land and sufficient water are vital for sustaining agriculture and livelihoods. Fertility of a soil is defined by its ability to provide all essential nutrients in adequate quantities and in the proper balance for the growth of plants – independent of direct application of nutrients – when other growth factors like light, temperature and water are favorable. This ability does not depend on the nutrient content of the soil only, but on its efficiency in transforming nutrients within the farm's nutrient circle.

In transformation of nutrients soil organisms play a key role. They break down biomass from crop residues, green manures and mulch and contribute to build up of soil organic matter, including humus, the soil's most important nutrient reservoir. They also play an essential role in transferring nutrients from the soil organic matter to the mineral stage, which is available to plants. Soil organisms also protect plants from disease and make the soil crumbly.

A fertile soil is easy to work, absorbs rainwater well, and is robust against siltation and erosion. It filters rain water and supplies us with clean drinking water. It neutralizes (buffers) acids, which pass through contaminated air to the soil surface, and decomposes pollutants such as pesticides rapidly. And last but not least a fertile soil is an efficient storage for nutrients and CO_2. In this way a fertile soil prevents the eutrophication of rivers, lakes and oceans and contributes to the reduction of global warming.

In the context of biological agriculture soil fertility is thus primarily the result of biological processes, not of chemical nutrients. A fertile soil is in active exchange with the plants, restructures itself and is capable of regeneration. The biological properties can be observed in the soil's conversion activity, in the presence and the visible traces of the organisms in it. The communities of microorganisms are robust and active at the right moment. In the self-regulating ecological equilibrium animals, plants and microorganisms all work for each other.

It is the responsibility of farmers to understand soil ecology to the point that they can create or restore the conditions for a robust balance in the soil. If a soil does not regularly bring good yields, farmers should investigate the reasons for it.

Properties of a fertile soil

A fertile soil:

- is rich in nutrients necessary for basic plant nutrition (including nitrogen, phosphorus, potassium, calcium, magnesium and sulphur);
- contains sufficient micronutrients for plant nutrition (including boron, copper, iron, zinc, manganese, chlorine and molybdenum);
- contains an appropriate amount of soil organic matter;
- has a pH in a suitable range for crop production (between 6.0 and 6.8);
- has a crumbly structure;
- is biologically active and
- has good water retention and supply qualities.

Essential plant nutrients

There are 16 essential nutrients that plants need in order to grow properly. Out of the 16 essential elements hydrogen, carbon and oxygen are obtained mainly from the air and from water. The other essential elements come from the soil and are generally managed by the farmers. Some of these nutrients are required in large amount in the plant tissues and are called macro (major) nutrients. Others are required in small amount and are called micro (minor) nutrients. Macronutrients include nitrogen (N), phosphorus (P) potassium (K), calcium (Ca), magnesium (Mg), and sulphur (S). Of these N, P and K are usually depleted from the soil first because plants need them in large amounts for their growth and survival, so they are known as primary nutrients. Ca, Mg and S are rarely limiting and are known as secondary nutrients. Where soils are acidic lime is often added, which contains large amounts of calcium and magnesium. Sulphur is usually found in sufficient amounts from the slowly decomposing soil organic matter. The micronutrients are boron (B), copper (Cu), iron (Fe), chloride (Cl), manganese (Mn), molybdenum (Mo), and zinc (Zn). Recycling organic matter such as crop residues and tree leaves is an excellent way of providing micronutrients to growing plants.

Soil organic matter

One of the most common objectives of organic farming—increased soil organic matter content—is difficult to measure.

When plant material and manure are mixed into the soil, they are decomposed and partly transformed into humus. Humus serves many purposes, for example:

- It acts as a reservoir of nutrients. The nutrients are released to the plants in a balanced way, which contributes to good plant health. Soil organic matter is the main nutrient pool for the plants beside nitrogen from symbiotic fixation.
- It increases the water holding capacity of the soil as it acts like a sponge with the ability to absorb and hold up to 90 % of its weight in water.

- It causes the soil to form strong complexes with clay particles, which improve soil structure and thus increase water infiltration, making the soil more resistant to erosion. Better soil structure also enhances root growth.
- Humus improves the exchange capacity for nutrients and avoids soil acidity.
- Soil biological activity is enhanced, which improves nutrient mobilisation from organic and mineral sources and the decomposition of toxic substances.
- Mycorrhizal colonisation is enhanced, which improves phosphorus supply.
- Compost has the potential to suppress soil borne pathogens, when applied to the soil.

Plant nutrition in organic farming relies on sound humus management. Proper management of soil organic matter requires some basic knowledge of the dynamics of soil organic matter. Aeration of the soil in combination with humidity and high temperature create favourable conditions for soil organisms and result in high biological activity enhancing decomposition of organic matter in the soil.

Under dry and cool conditions soil biological activity is strongly reduced resulting in a reduction or even in a standstill of transformation processes. Managing soil organic matter for plant nutrition and soil organic matter level means knowing, when and how to manage temperature, oxygen and moisture conditions of the soil and interfering (or not interfering) to stimulate or calm down decomposition and build-up of soil organic matter. Excessive tillage for example stimulates decomposition of soil organic matter, whereas cooling the soil with a soil cover slows it down.

Building soil organic matter is a long-term process, but investing into it is highly beneficial to crop or forage production, and contributes to higher and more reliable yields. There are different ways of maintaining or improving soil organic matter:

- Growing green manure, mostly legumes, for the amount of biomass they build. Before flowering they are cut and worked into the soil.

- Intercropping cover crops such as velvet bean, Tithonia, lablab and others as living mulch. The cover crop is regularly slashed, when it competes too much with the main crop.
- Mulching with especially hard-to-compost or woody materials, like dry crop residues or green manure crops, which have been grown to maturity, also can on tribute to a slow increase of soil organic matter over time.
- Trees and shrubs for agroforestry can be grown in the fields with crops where they are regularly pruned and the branches are used as mulch. They may also be planted on the edges of a field or on fallow plots.
- Residues from harvested crops in the form of husks, leaves, roots, peelings, branches and twigs should be returned to the fields either as compost, as mulching materials, or for incorporation into the soil.
- Depending on the financial situation of the farm, additional materials from agro-processing like wood shavings, or coffee or rice husks, or from food industry like seed cakes can be purchased.
- Integration of livestock helps to quickly improve soil organic matter, when livestock excreta and bedding are properly recycled.

The amount and the quality of organic matter supplied to the soil influence the content of organic matter in the soil. A regular supply of organic matter provides the best conditions for balanced plant nutrition. Estimates say that in humid tropical climates 8.5 tons, in sub-humid climate 4 tons, and in semi-arid 2 tons of biomass are needed per hectare and year to maintain soil carbon levels of 2, 1 and 0.5 %, respectively.

Burning organic residues and standing dead biomass (such as crops left on a field) is a crime to the environment! All the benefits that may be derived from incorporating organic matter are lost and, if the plant material is burned, the atmosphere is polluted. The ashes contain nutrients that are directly available to the plants; however, large amounts of carbon, nitrogen and sulphur are released as gas and are lost. The nutrients in the ash are also easily washed out with the first rain. The burning also harms beneficial insects and soil organisms.

Identifying sources of biomass

The majority of the farmers by far do not exploit the potentials of on-farm production of organic soil inputs. Realizing the potentials of farm-own resources can be essential for long-term sustainability of the farm, as it helps to reduce the cost of buying organic materials like manure or mineral fertilizers. Instead of buying farm inputs the farmers may use part of the savings for buying seeds for green manures and feed plants or for purchasing own livestock.

Due to limited landholdings and lack of livestock, some of the farmers may be unable to produce adequate quantities of green manure and compost. These farmers will depend on outside sources of organic materials to maintain the fertility of their soils.

Feed the Soil Approach

Many organic farmers prefer to base fertility management on a *feed the soil* approach. The purpose of this approach is twofold. When organic nutrients are added to the soil, microbial activity increases. In this sense, organic farmers are "feeding the microbes." Increased microbial activity improves soil physical properties. For example, when microbial activity increases, soil tilth improves. In addition, microbial activity speeds nutrient cycling, increasing the availability of nutrients for plant uptake (when mineralization exceeds immobilization by microbes).

Improvement in agricultural sustainability requires, alongside effective water and crop management, the optimal use and management of soil fertility and soil physical properties. Both rely on soil biological processes and soil biodiversity. This calls for the widespread adoption of management practices that enhance soil biological activity and thereby build up long-term soil productivity and health.

The major steps to enhance soil fertility are:

- Conservation tillage
- Composting and Vermicomposting
- Mixed Cropping
- Crop rotation and varietal selection
- Green manuring
- Farmyard manure
- Oil seed refuge
- Leguminous plants
- Biofertilisers
- Crop residue recycling
- Panchagavya as foliar spray
- Effective mircobes technology

Manures

Manures are plant and animal wastes that are used as sources of plant nutrients. They release nutrients after their decomposition. The art of collecting and using wastes from animal, human and vegetable sources for improving crop productivity is as old as agriculture. Manures are the organic materials derived from animal, human and plant residues which contain plant nutrients in complex organic forms. Naturally occurring or synthetic chemicals containing plant nutrients are called fertilizers. Manures with low nutrient, content per unit quantity have longer residual effect besides improving soil physical properties compared to fertilizer with high nutrient content. Major sources of manures are:

1. Cattle shed wastes-dung, urine and slurry from biogas plants
2. Human habitation wastes-night soil, human urine, town refuse, sewage, sludge and sullage
3. Poultry Jitter, droppings of sheep and goat
4. Slaughterhouse wastes-bone meal, meat meal, blood meal, horn and hoof meal, Fish wastes

5. Byproducts of agro industries-oil cakes, bagasse and press mud, fruit and vegetable processing wastes etc
6. Crop wastes-sugarcane trash, stubbles and other related material
7. Water hyacinth, weeds and tank silt, and
8. Green manure crops and green leaf manuring material

Manures can also be grouped, into bulky organic manures and concentrated organic manures based on concentration of the nutrients.

Farmyard manure

Farmyard manure commonly describes a more or less decomposed mixture of livestock dung and urine (mostly from cattle) mixed with straw and litter, which was used as bedding material. It may also contain residues from the fodder fed to the cattle and decomposed household waste.

Farmyard manure is extremely valuable organic manure. Farmyard manure contains large amounts of nutrients. The availability of phosphorus and potassium from farmyard manure is similar to that of chemical fertilizers. Chicken manure is rich in phosphorus. When dung and urine from cattle are mixed, they form a well-balanced source of nutrients for plants.

Many farmers still underestimate the value of animal manure. In many places, it is dried and burned for cooking or just not recognized as a source of nutrients and organic matter. By drying or burning farmyard manure, large quantities of organic matter and nutrients are lost from agricultural systems. Appropriate recycling of nutrients on the farm, especially if it comes from a high-value source, is a principle of organic farming. Therefore, proper handling and use of animal manures are essential to ensure that the nutrients in the manure are preserved and the risks of causing environmental pollution are minimized.

Most farmers do not own animals, and neither do they have access to animal manure. Growing animal feed and integrating livestock into the farm not only provides milk and or meat and other animal products, but also some animal manure In areas with mixed crop-livestock farming systems manure is likely to be available to most households, although at varying levels.

Improving the value of farm yard manure

Farmyard manure refers to the decomposed mixture of dung and urine of farm animals along with litter and left over material from roughages or fodder fed to the cattle. On an average well decomposed farmyard manure contains 0.5 per cent N, 0.2 per cent P_2O_5 and .0.5 per cent K_2O. The present method of preparing farmyard manure by the farmers is defective. Urine, which is wasted, contains one per cent nitrogen and 1.35 per cent potassium. Nitrogen present in urine is mostly in the form of urea which is subjected to volatilization losses. Even during storage, nutrients are lost due to leaching and volatilization. However, it is practically impossible to avoid losses altogether, but can be reduced by following improved method of preparation of farmyard manure. Trenches of size 6 m to 7.5 m length, 1.5 m to 2.0 m width and 1.0 m deep are dug.

All available litter and refuse is mixed with soil and spread in the shed so as to absorb urine. The next morning, urine soaked refuse along with dung is collected and placed in the trench. A section of the trench from one end should be taken up for filling with daily collection. When the section is filled up to a height of 45 cm to 60 cm above the ground level, the top of the heap is made into a dome and plastered with cow dung earth slurry. The process is continued and when the first trench is completely filled, second trench is prepared.

The manure becomes ready for use in about four to five months after plastering. If urine is not collected in the bedding, it can be collected along with washings of the cattle shed in a cemented pit from which it is later added to the farmyard manure pit.

Partially rotted farmyard manure has to be applied three to four weeks before sowing while well rotted manure can be applied immediately before sowing. Generally 10 to 20 t/ha is applied, but more than 20 t/ ha is applied to fodder grasses and vegetables. In such cases farmyard manure should be applied at least 15 days in advance to avoid immobilization of nitrogen. The existing practice of leaving manure in small heaps scattered in the field for a very long period leads to loss of nutrients. These losses can be reduced by spreading the manure and incorporating by ploughing immediately after application.

Vegetable crops like potato, tomato, sweet-potato, carrot, raddish, onion etc., respond well to the farmyard manure. The other responsive crops are sugarcane, rice, napier grass and orchard crops like oranges, banana, mango and plantation crop like coconut.

The entire amount of nutrients present in farmyard manure is not available to the crop immediately. About 30 per cent of nitrogen, 60 to 70 per cent of phosphorus and 70 per cent of potassium are available to the first crop.

Sheep and Goat Manure

The droppings of sheep and goats contain higher nutrients than farmyard manure and compost. On an average, the manure contains 3 per cent N, 1 per cent P_2O_5 and 2 per cent K_2O.It is applied to the field in two ways. The sweeping of sheep or goat sheds are placed in pits for decomposition and it is applied later to the field. The nutrients present in the urine are *wasted* in this method. The second method is sheep penning, wherein sheep and goats are kept overnight in the field and urine and faecal matter added to the soil is incorporated to a shallow depth by working blade harrow or cultivator or cultivator.

Poultry Manure

The excreta of birds ferment very quickly. If left exposed, 50 percent of its nitrogen is lost within 30 days. Poultry manure contains higher nitrogen and phosphorus compared to other bulky organic manures. The average nutrient content is 3.03 per cent N; 2.63 per cent P_2O_5 and 1.4 per cent K_2O.

Concentrated organic manures

Concentrated organic manures have higher nutrient content than bulky organic manure. The important concentrated organic manures are oilcakes, blood meal, fish manure etc. These are also known as organic nitrogen fertilizer. Before their organic nitrogen is used by the crops, it is converted through bacterial action into readily usable ammoniacal nitrogen and nitrate nitrogen. These organic fertilizers are, therefore, relatively slow acting, but they supply available nitrogen for a longer period.

Oil cakes

After oil is extracted from oilseeds, the remaining solid portion is dried as cake which can, be used as manure. The oil cakes are of two types:

- Edible oil cakes which can be safely fed to livestock; e.g.: Groundnut cake, Coconut cake etc., and
- Non edible oil cakes which are not fit for feeding livestock; e.g.: Castor cake, Neem cake, Mahua cake etc.,

Both edible and non-edible oil cakes can be used as manures. However, edible oil cakes are fed to cattle and non-edible oil cakes are used as manures especially for horticultural crops. Nutrients present in oil cakes, after mineralization, are made available to crops 7 to 10 days after application. Oilcakes need to be well powdered before application for even distribution and quicker decomposition.

Other Concentrated Organic Manures

Blood meal when dried and powdered can be used as manure. The meat of dead animals is dried and converted into meat meal which is a good source of nitrogen. Average nutrient content of animal based concentrated organic manures is given as follows.

Compost

Compost is a common name used for plant and animal material (mainly animal manure) that has been fully decomposed in a targeted process initialized and controlled by man. Compared with uncontrolled decomposition of organic material as it naturally occurs, decomposition in the composting process occurs at a faster rate, reaches higher temperatures and results in a product of higher quality.

Composting is a process where microorganisms break down organic matter to produce humus like substance called compost. The process occurs naturally provided the right organisms, water, oxygen, organic material and nutrients are available for microbial growth. By controlling these factors, the composting process can occur at a much faster rate.

Composting is a means of ensuring or improving long-term soil fertility, especially to smallholder farmers with no or little access to manures and fertilizers. Compost is more than a fertilizer. It is not just a nutrient source, but also acts on the structure of the soil and on its capacity to hold and provide nutrients and water. Its main value lies in its long-term effect on soil fertility.

Compost contributes to an increase of the organic matter content of the soil and thus to a better soil structure. It clearly enhances drought resistance of crops.

During the composting process diseases, pests and weed seeds are destroyed. Even viruses are destroyed, if a high temperature is reached. Thus, composting helps solve common problems associated to the management of plant residues.

Compost also increases biological activity of the soil and its capacity to positively influence biological control of root rot diseases from fungi, bacteria and nematodes.

In the composting process nutrients are adsorbed into the organic matter, microorganisms and humus. The humic substances are relatively resistant to microbial decomposition. Thus, the nutrients are released slowly and are not easily lost.

Why composting is necessary?

- The rejected biological materials contain complex chemical compounds such as lignin, cellulose, hemicellulose, polysaccharides, proteins, lipids etc.
- These complex materials cannot be used as such as resource materials.
- The complex materials should be converted into simple inorganic element as available nutrient.
- The material put into soil without conversion will undergo conversion inside the soil.
- This conversion process takes away all energy and available nutrients from the soil affecting the crop.
- Hence conversion period is mandatory.

Advantages of Composting

- Volume reduction of waste
- Final weight of compost is very less
- Composting temperature kill pathogen, weed seeds and seeds
- Matured compost comes into equilibrium with the soil
- During composting number of wastes from several sources are blended together
- Excellent soil conditioner
- Saleable product
- Improves manure handling
- Reduces the risk of pollution
- Pathogen reduction
- Additional revenue
- Suppress plant diseases and pests
- Reduce or eliminate the need for chemical fertilizers
- Promote higher yields of agricultural crops
- Facilitate reforestation, wetlands restoration, and habitat revitalization efforts by amending contaminated, compacted, and marginal soils
- Cost-effectively remediate soils contaminated by hazardous waste
- Remove solids, oil, grease, and heavy metals from stormwater runoff
- Capture and destroy 99.6 percent of industrial volatile organic chemicals (VOCs) in contaminated air
- Provide cost savings of at least 50 percent over conventional soil, water, and air pollution remediation technologies, where applicable

The Phases of Composting Process

There are three main phases of composting process

- The heating phase: In this phase the temperature of the compost heap rises to 60° to 70°C. The decomposition of the materials used in compost pits occurs during the heating phase. Bacteria are mainly active during the first phase. The heat destroys diseases, pests, weeds, and roots.
- The cooling phase: The second phase of composting is cooling phase in which the temperature in the compost heap declines slowly and will remain 25-45°C. In this phase fungi start the decomposition of straw, fibers and wooden material.
- The maturing phase: During this phase nutrients are mineralized and humic acids and antibiotics are built up. At the end of this phase the compost needs much less water than in the heating phase.

Methods of composting

Coimbatore method: In this method of composting is done in pits of different sizes depending on the waste material available. A layer of waste materials is first laid in the pit. It is moistened with a suspension of 5-10 kg cow dung in 2.5 to 5.0 I of water and 0.5 to 1.0 kg fine bone meal sprinkled over it uniformly. Similar layers are laid one over the other till the material rises 0.75 m above the ground level. It is finally plastered with wet mud and left undisturbed for 8 to 10 weeks. Plaster is then removed, material moistened with water, given a turning and made into a rectangular heap under a shade. It is left undisturbed till its use.

Indore method: In this method of composting, organic wastes are spread in the cattle shed to serve as bedding. Urine soaked material along with dung is removed every day and formed into a layer of about 15 cm thick at suitable sites. Urine soaked earth, scraped from cattle sheds is mixed with water and sprinkled over the layer of wastes twice or thrice a day. Layering process continued for about a fortnight. A thin layer of well decomposed compost is sprinkled over top and the heap given a turning and reformed. Old compost acts as inoculum for decomposing the material. The heap is left undisturbed for about a month. Then it is thoroughly moistened and given a turning. The compost is ready for application in another month.

Bangalore method: In this method of composting, dry waste material of 25 cm thick is spread in a pit and a thick suspension of cow dung in water is sprinkled over for moistening. A thin layer of dry waste is laid over the moistened layer. The pit is filled alternately with dry layers of material and cow dung suspension till it rises 0.5 m above ground level. It is left exposed without covering for 15 days. It is given a turning, plastered with wet mud and left undisturbed for about 5 months or till required.

In Coimbatore method, there is anaerobic decomposition to start with, following by aerobic fermentation. It is the reverse in Bangalore method. The Bangalore compost is not so thoroughly decomposed as the Indore compost or even as much as the Coimbatore compost, but it is bulkiest.

Compost is a rich source of organic matter. Soil organic matter plays an important role in sustaining soil fertility, and hence in sustainable agricultural production. In addition to being a source of plant nutrient, it improves the physico-chemical and biological properties of the soil. As a result of these improvements, the soil:

i. Becomes more resistant to stresses such as drought, diseases and toxicity;

ii. Helps the crop in improved uptake of plant nutrients; and

iii. Possesses an active nutrient cycling capacity because of vigorous microbial activity.

These advantages manifest themselves in reduced cropping risks, higher yields and lower outlays on inorganic fertilizers for farmers.

The Benefits of Using Composts to Agriculture

Compost has been considered as a valuable soil amendment for centuries. Most people are aware that using composts is an effective way to increase healthy plant production, help save money, reduce the use of chemical fertilizers, and conserve natural resources.

Compost provides a stable organic matter that improves the physical, chemical, and biological properties of soils, thereby enhancing soil quality and crop production. When correctly applied, compost has the following beneficial effects on soil properties, thus creating suitable conditions for root development and consequently promoting higher yield and higher quality of crops.

A. Improves the Physical Properties of Soils

- Reduces the soil bulk density and improves the soil structure directly by loosening heavy soils with organic matter, and indirectly by means of aggregate-stabilizing humus contained in composts. Incorporating composts into compacted soils improves root penetration and turf establishment.
- Increases the water-holding capacity of the soil directly by binding water to organic matter, and indirectly by improving the soil structure, thus improving the absorption and movement of water into the soil. Therefore, water requirement and irrigation will be reduced.
- Protects the surface soil from water and wind erosion by reducing the soil-dispersion action of beating raindrops, increasing infiltration, reducing water runoff, and increasing surface wetness. Preventing erosion is essential for protecting waterways and maintaining the quality and productivity of the soil.
- Helps bind the soil particles into crumbs by the fungi or actinomycetes mycelium contained in the compost and stimulated in the soil by its application, generally increasing the stability of the soil against wind and water erosion.
- Improves soil aeration and thus supplies enough oxygen to the roots and escapes excess carbon dioxide from the root space.
- Increases the soil temperature directly by its dark color, which increases heat absorption by the soil, and indirectly by the improved soil structure.
- Helps moderate soil temperature and prevents rapid fluctuations of soil temperature, hence, providing a better environment for

root growth. This is especially true of compost used as a surface mulch.

B. Enhances the Chemical Properties of Soils

- Enables soils to hold more plant nutrients and increases the cation exchange capacity (CEC), anion exchange capacity (AEC), and buffering capacity of soils for longer periods of time after composts are applied to soils. This is important mainly for soils containing little clay and organic matter.
- Builds up nutrients in the soil. Composts contain the major nutrients required by all plants [N,P,K, calcium (Ca), magnesium(Mg), and S] plus essential micronutrients or trace elements, such as copper (Cu), zinc (Zn), iron (Fe), manganese (Mn), boron (B), and molybdenum (Mb).
- The nutrients from mature composts are released to the plants slowly and steadily. The benefits will last for more than one season.
- Stabilizes the volatile nitrogen of raw materials into large protein particles during composting, thereby reducing N losses.
- Provides active agents, such as growth substances, which may be beneficial mainly to germinating plants.
- Adds organic matter and humus to regenerate poor soils.
- Buffers the soil against rapid changes due to acidity, alkalinity, salinity, pesticides, and toxic heavy metals.

C. Improves the Biological Properties of Soils

- Supplies food and encourages the growth of beneficial microorganisms and earthworms.
- Helps suppress certain plant diseases, soil borne diseases, and parasites.
- Research has shown that composts can help control plant diseases (e.g. Pythium root rot, Rhizoctonia root rot, chili wilt, and parasitic nematode) and reduce crop losses. A major California fruit and vegetable grower was able to cut pesticide use by 80% after three years of compost applications as part of

an organic matter management system. Research has also indicated that some composts, particularly those prepared from tree barks, release chemicals that inhibit some plant pathogens. Disease control with compost has been attributed to four possible mechanisms:

- Successful competition for nutrients by beneficial microorganisms;
- Antibiotic production by beneficial microorganisms;
- Successful predation against pathogens by beneficial microorganisms;
- Activation of disease-resistant genes in plants by composts; and
- High temperatures that result from composting kill pathogens.
- Reduces and kills weed seeds by a combination of factors including the heat of the compost pile, rotting, and premature germination.

Economic and Social Benefits of Composting

The economic and social benefits of composting include the following:

- Brings higher prices for organically grown crops.
- Composting can offer several potential economic benefits to communities:
- Extends current landfill longevity and delays the construction of a more expensive replacement landfill or incinerator.
- Reduces or avoids landfill or combustor tipping fees, and reduces waste disposal fees and long-distance transportation costs.
- Offers environmental benefits from reduced landfill and combustion use.
- Creates new jobs for citizens.
- Produces marketable products and a less-cost alternative to standard landfill cover, artificial soil amendments, and conventional bioremediation techniques.

- Provides a source of plant nutrients and improves soil fertility; results in significant cost savings by reducing the need for water, pesticides, fungicides, herbicides, and nematodes.
- Used as an alternative to natural topsoil in new construction, landscape renovations, and container gardens. Using composts in these types of applications is not only less expensive than purchasing topsoil, but it can also often produce better results when establishing a healthy vegetative cover.
- Used as mulch for trees, orchards, landscapes, lawns, gardens, and makes an excellent potting mix. Placed over the roots of plants, compost mulch conserves water and stabilizes soil temperatures. In addition, it keeps plants healthy by controlling weeds, providing a slow release of nutrients, and preventing soil loss through erosion.

Vermicomposting

Vermicomposting is the process of turning organic debris into worm castings. The worm castings are very important to the fertility of the soil. The castings contain high amounts of nitrogen, potassium, phosphorus, calcium, and magnesium. Several researchers have demonstrated that earthworm castings have excellent aeration, porosity, structure, drainage, and moisture-holding capacity. The content of the earthworm castings, along with the natural tillage by the worms burrowing action, enhances the permeability of water in the soil. Worm castings can hold close to nine times their weight in water.

In the 1996 Summer Olympics in Sydney, Australia, the Australians used worms to take care of their tons and tons of waste. They then found that waste produced by the worms was could be very beneficial to their plants and soil. People in the U.S. have commercial vermicomposting facilities, where they raise worms and sell the castings that the worms produce.

Earthworms have been on the Earth for over 20 million years. In this time they have faithfully done their part to keep the cycle of life continuously moving. Their purpose is simple but very important. They are nature's way of recycling organic nutrients from dead tissues back to living organisms. Many have recognized the value of these worms. Ancient civilizations, including Greece and Egypt valued the role earthworms played in soil. The Egyptian Pharaoh, Cleopatra said, "Earthworms are sacred". She recognized the important role the worms played in fertilizing the Nile Valley croplands after annual floods. Charles Darwin was intrigued by the worms and studied them for 39 years. Referring to an earthworm, Darwin said, "It may be doubted whether there are many other animals in the world which have played so important a part in the history of the world. The earthworm is a natural resource of fertility and life."

Materials for preparation of vermicompost

Any types of biodegradable wastes viz.,

1. Crop residues
2. Weed biomass
3. Vegetable waste
4. Leaf litter
5. Biodegradable Hostel refuses
6. Waste from agro-industries
7. Biodegradable portion of urban and rural wastes
8. Earthworms

Selection of suitable earthworm

For vermicompost production, the surface dwelling earthworm alone should be used. The earthworm, which lives below the soil, is not suitable for vermicompost production. The African earthworm (*Eudrillus eugeniae*), Red worms (*Eisenia foetida*) and composting worm (*Perionyx excavatus*) are promising worms used for vermicompost production. All the three worms can be mixed together for vermicompost production. The African worm (*Eudrillus eugeniae*) is preferred over other two types, because it produces higher production of vermicompost in short period of time and younger ones in the composting period.

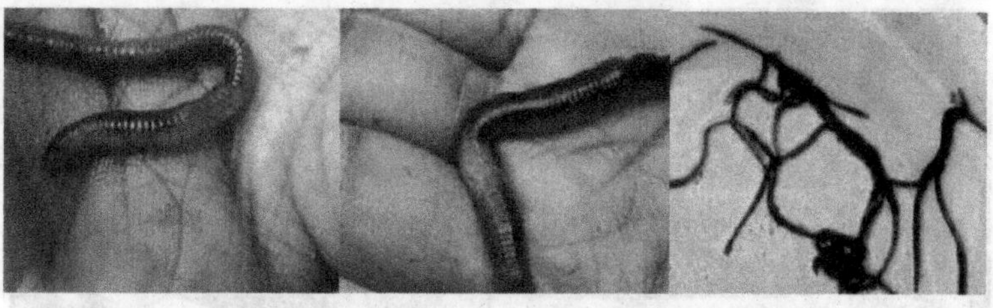

African earthworm (*Eudrillus euginiae*) — Tiger worm or Red wrinkle (*Eisenia foetida*) — Asian worms (*Perinonyx ecavatus*)

Fig. 2. Suitable earthworms for vermicompost preparation

Steps for vermicompost preparation

Select a thatched roof or place with shade, high humidity and cool

↓

A cement tub may be constructed to a height of 2½ feet and a breadth of 3 feet or over the hand floor, hollow blocks / bricks may be arranged in compartment to a height of one feet, breadth of 3 feet and length to a desired level.

↓

Worm bed (3 cm) - by placing after saw dust or husk or coir waste or sugarcane trash in the bottom of tub / container

Moistened with water ↓

A layer of fine sand (3 cm) should be spread over the culture bed followed by a layer of garden soil (3 cm).

↓

Moistened with water

Keep it all for predigestion (frequent turnings) for 20 days

The predigested waste material should be mud with 30% cattle dung either by weight or volume.

↓

Maintained at 60% moisture level ↓

↓

1 kg of worm (1000 Nos.) is released. Water should be sprinkled over the bed rather than pouring the water.

↓

The castings formed on the top layer are collected periodically

Thatched house

Introduction of earthworm

Coir pith composting

The largest by products of coconut is coconut husk from which coir fibre is extracted. This extraction process generates a large quantity of dusty material called coir dust or coir pith. Large quantity of coir waste of about 7.5 million tonnes is available annually form coir industries in India. In Tamil Nadu, 5 lakh tonnes of coir dust is available at present.

Coir pith is collected from the coir industry without any fibre. If fibrous materials are present, it is removed by sieving at the source itself.

It is better to have an elevated shady place for composting. In between coconut trees, shade under any tree is good for composting.

Coir pith composting is an aerobic process. So it should be heaped above the soil. Coir pith should be spread to the length of 4 feet and breadth of 3 feet. Initially coir pith should be put up for 3 inch height and thoroughly moistened.

After moistening, nitrogenous source material should be added. The nitrogenous source may be in the form of urea (5 kg/ton) or fresh poultry litter (200 kg/ton). This 5 kg equally divided into five portions and in alternative layer of coir pith one kg of urea should be applied.

One has to proportionally divide and put the required amount of poultry litter over the coir pith. For example if one ton coir pith is divided into 10 portions, in the first layer, 100 kg poultry litter is added.

After adding, the nitrogen source, the microbial inoculums *Pleurotus* and TNAU biomineralizer (2%) are added over the material. Over this one portion of coir pith is added and the same input mentioned above should be added. It is advisable to make a heap up to minimum of 4 feet height. But beyond 5 feet, it requires machinery to handle the materials.

The compost heap should be turned once in 10 days to allow the stale air trapped inside the compost material to go out and fresh air will get in. The other way of giving aeration is inserting perforated unused PVC or iron pipe in the composting material both vertically and horizontally.

Maintaining optimum moisture (60 % moisture) is the prerequisite for uniform composting of waste material.

After 60 days the composted material which is obtained from sieving is ready for use. It is recommended that 5 tons of composted coir pith per hectare;

Fig. 3. Coir pith compost

Sugarcane trash composting

Sugarcane produces about 10 to 12 tonnes of dry leaves per hectare per crop. The detrashing is done on 5th and 7th month during its growth period. The sugarcane trash incorporation in the soil reduces soil EC, improves the water holding capacity, better soil aggregation and thereby improves porosity in the soil. Sugarcane trash incorporation reduces the bulk density of the soil and there is an increase in infiltration rate and decrease in penetration resistance. The direct incorporation of chopped trash increases the availability of nutrients leading to soil fertility.

Collection of trash

The detrashed material has to be pooled together and transported to the compost yard. If no compost yard is available to farmer, anyone of the corner area in the sugarcane field itself can be used for making composting. Sugarcane trash is lengthy one. It is recommended to shred the waste into small particles. This process reduces the volume of material, increases the surface area of the waste. If the waste material contains more surface area, more microorganisms work effectively on the surface and degradation will be faster. Shredder is the ideal instrument to shred all the sugarcane trash. Chop cutter machine can also be used. If no machinery is available manual shredding is recommended.

TNAU biomineralizer

TNAU biomineralizer is the consortium of microorganism recommended for composting all the agro wastes. For one ton of trash, two kg inoculums are recommended. Without the inoculation of microbial consortium, the composting process will take its own time. Animal dung or fresh poultry litter can be used as a source of nitrogen to reduce the C: N ratio. For one ton of sugarcane trash 50 kg fresh dung is recommended. The dung can be mixed with 100 liters of water and thoroughly mixed with sugarcane trash. Rock phosphate at 5 kg I ton waste can be added to increase the phosphorus content of the compost.

After mixing all the inputs with sugarcane trash, heap should be formed with a minimum height of 4 feet. This height is required to generate more heat in the composting process, and the generated heat will be retained long time inside the material.

The compost material should be turned periodically once in 15 days to allow more aeration inside the material. In the turning process, bottom layer comes to top and top layer comes to bottom. So that uniform composting will occur.

Throughout the composting period about 60% moisture should be maintained. If composting material is allowed to dry, all the established microorganisms get killed and composting process will be terminated.

Volume reduction, earthy odor, brownish black colour and reduction in particle size are important parameters to be observed for assessing compost maturity. Once the compost attained the maturity, the compost heap should be disturbed and spread the material for curing. After 24 hrs the composted material can be sieved through 4 mm sieve to get uniform compost material. The residues available after sieving will be recycled to the next composting batch for further composting.

The enriched compost can be applied at the rate of 5 tons per hectare, as basal application to the field. Whatever the compost derived from sugarcane trash, it can be ploughed back into sugarcane field to enrich the soil.

Crop residue composting

Crop residues are the non-economic plant parts that are left in the field after harvest. The harvest refuses include straws, stubble, stover and haulms of different crops. Crop remains are also from thrashing sheds or that are discarded during crop processing. This includes process wastes like groundnut shell, oil cakes, rice husks and cobs of maize, sorghum and *cumbu*. The greatest potential as a biomass resource appears to be from the field residues of sorghum, maize, soybean, cotton, sugarcane etc. In Tamil Nadu, at present 190 lakh tones of crop residues are available for use.

Shred the waste into small particles. This process reduces the volume of material, increases the surface area of the waste. Carbon and nitrogen ratio decides the initiation of composting process.

Shredding of waste

Shredded pieces

Green coloured waste materials like *Glyricidia* leaves, *Parthenium*, freshly harvested weeds; *Sesbania* leaves are rich in nitrogen, whereas brown coloured waste material like straw, coir dust, dried leaves and dried grasses are rich in carbon. Animal dung is also a good source of nitrogen. While making heap formation, alternative layers of carbon rich material, animal dung and nitrogen rich material are to be heaped to get a quicker result in composting.

Minimum 4 feet height at an elevated place to have a sufficient shade should be maintained for composting. While heap formation, all the crop residues should be mixed in an alternate layers of carbon and nitrogen rich material with intermittent layers of animal dung are essential. After heap formation the material should be thoroughly moistened.

Copost sieving

For one ton of crop wastes 2 kg of TNAU Biomineralizer is recommended. This two kg Biomineralizer should be mixed with 20 liters of water and made slurry. When the compost heap is formed in between layers the slurry should be inoculated, so that it mixes with the waste material thoroughly for uniform coating of microorganism on the waste material.

Sufficient quantity of oxygen should be available inside the compost heap. Normally to allow the fresh air to get inside, the compost heap should be turned upside down, once in fifteen days. Throughout the composting period about 60% moisture should be maintained.

Volume reductions, black colour, earthy odor, reduction in particle size are all the physical factors to be observed for compost maturity. After satisfying with the compost maturity index, the compost heap can be disturbed and spread on the floor for curing. After curing for one day, the composted material is sieved through 4 mm sieve to get uniform composted material. The residues collected after composting has to be again composted to finish the composting process.

Compost enrichment

The harvested compost should be heaped in a shade, preferably on a hard floor. The beneficial microorganisms like Azotobacter or Azospirillum, Pseudomonas, Phosphobacteria (0.2%) and rock phosphate (2%) have to be inoculated for one ton of compost. About 40 % moisture should be maintained for the maximum growth of inoculated microorganism. This incubation should be allowed for 20 days for the organism to reach the maximum population. Now the compost is called as enriched compost. The advantage of enriched compost over normal compost is the quality manure with higher nutrient status with high number of beneficial microorganisms and plant growth promoting substances.

For one hectare of land 5 tons of enriched biocompost is recommended. It can be used as basal application in the field before sowing/planting.

Tricho-composting

A composting method developed by central Cotton Research Institute, Nagpur. Composting of Cotton Stalks of five ha in a pit of 6 x 3 x 1m using the cellulytic fungis *Trichoderma viride* is worth mentioning.

Conversion of solid waste with the help of microorganisms like *Trichoderma viride* needs no heavy equipment or investment. All work is proposed to be done by hands, which are available in plenty in all villages provided they are paid a reasonable wage.

All biodegradable materials, crop dry wastes are not suitable as feed for the livestock such as crop residues of cotton, pigeonpea, sorghum, sugarcane all weeds including (*Parthenium hysterophorus*) is chopped into small pieces of say 3-5 cms length by hand or chaff cutter or a threshing machine.

These small pieces of material can then be soaked in water, for a few minutes and then laid on the site in a raised bed of 2 x 3 m size with a height of only 30 cm.

The mixture of compost making micro- organisms like *Trichoderma viride* is prepared with 200 litres of water, 30 kg cow dung and 30 kg soil. About 40 litres of sprinkled on the raised heap.

Another layer of the raw material is then laid on the first layer and again the same mixture is sprinkled. The process is repeated for the next 3 layers.

The total height of the heap will be about 2 m on the first day. The heap is then covered by mud or grass.

The micro organisms become active on the substrate immediately and in the process they themselves also multiply.

If culture is not available about 100 kg of the prepared compost could be used as a culture.

In about 6 week's time, the compost will be ready after turned upside down. If needed compost can be sieved through a mesh of 2.5 x 2.5 cm size and transported to the field loose or in bags.

Approximately the total cost of production will be Rs. 1600/ ton of compost.

With the adoption of organic farming by a large majority of farmers in the years to come demand for all organic manures is bound to increase phenomenally. The farmers are not willing to exert themselves and they need something, ready to use like bags of urea or bottles of pesticide: Therefore it can be safely assumed that such production activity will flourish in any farm / village.

Biofertilisers

Biofertilizers are defined as preparations containing living cells or latent cells of efficient strains of microorganisms that help crop plants' uptake of nutrients by their interactions in the rhizosphere when applied through seed or soil. They accelerate certain microbial processes in the soil which augment the extent of availability of nutrients in a form easily assimilated by plants.Plants have a number of relationships with fungi, bacteria, and algae, the most common of which are with mycorrhizae, rhizobium, and cyanophyceae etc. These are known to deliver a number of benefits including plant nutrition, disease resistance,and tolerance to adverse soil and climatic conditions.

Use of biofertilizers is one of the important components of integrated nutrient management, as they are cost effective and renewable source of plant nutrients to supplement the chemical fertilizers for sustainable agriculture. Several microorganisms and their association with crop plants are being exploited in the production of biofertilizers. They can be grouped in different ways based on their nature and function.

Table 6. Classification of bio fertilisers

S. No.	Groups	Examples
N2 fixing Biofertilizers		
1.	Free-living	*Azotobacter, Beijerinkia, Clostridium, Klebsiella, Anabaena, Nostoc,*
2.	Symbiotic	*Rhizobium, Frankia, Anabaena azollae*
3.	Associative Symbiotic	*Azospirillum*
P Solubilizing Biofertilizers		
1.	Bacteria	*Bacillus megaterium* var. *phosphaticum, Bacillus subtilisBacillus circulans, Pseudomonas striata*
2.	Fungi	*Penicillium* sp, *Aspergillus awamori*
P Mobilizing Biofertilizers		
1.	Arbuscular mycorrhiza	*Glomus* sp.,*Gigaspora* sp.,*Acaulospora* sp., *Scutellospora* sp. & *Sclerocystis* sp.
2.	Ectomycorrhiza	*Laccaria* sp., *Pisolithus* sp., *Boletus* sp., *Amanita* sp.
3.	Ericoid mycorrhizae	*Pezizella ericae*
4.	Orchid mycorrhiza	*Rhizoctonia solani*

(Contd.)

Biofertilizers for Micro nutrients	
1.Silicate and Zinc solubilizers	*Bacillus* sp.
Plant Growth Promoting Rhizobacteria	
1.Pseudomonas	*Pseudomonas fluorescens*

Working Principles of Bio-fertilizer

- Biofertilizers fixed atmospheric nitrogen in the soil and root nodules of legume crops and make it available to the plant.
- They solubilise the insoluble forms of phosphates like tricalcium, iron and aluminium phosphates into available forms.
- They scavenge phosphate from soil layers.
- They produce hormones and anti metabolites which promote root growth.
- They decompose organic matter and help in mineralization in soil.
- When applied to seed or soil, biofertilizers increase the availability of nutrients and improve the yield by 10 to 25% without adversely affecting the soil and environment.

Application of Biofertilizers

1. Seed treatment or seed inoculation
2. Seedling root dip
3. Main field application

Seed treatment

One packet of the inoculant is mixed with 200 ml of rice kanji to make slurry. The seeds required for an acre are mixed in the slurry so as to have a uniform coating of the inoculant over the seeds and then shade dried for 30 minutes. The shade dried seeds should be sown within 24 hours. One packet of the inoculant (200 g) is sufficient to treat 10 kg of seeds.

Seedling root dip

This method is used for transplanted crops. Two packets of the inoculant are mixed in 40 litres of water. The root portion of the seedlings required for an acre is dipped in the mixture for 5 to 10 minutes and then transplanted.

Main field application

Four packets of the inoculant is mixed with 20 kgs of dried and powdered farm yard manure and then broadcasted in one acre of main field just before transplanting.

A. Nitrogen fixers

Rhizobium spp.

These are gram positive soil bacteria and symbiotic nitrogen fixer which assimilates atmospheric nitrogen and fixes in the root nodule, formed in the roots of leguminous plants and also in some nonleguminous plants. The root nodulating rhizobia are as:

R .leguminosarum nodulates pea; *R . phaseoli* nodulates bean and *Bradyrhizobium* nodulates soybean. But *Azorhizobium caulinodans* is one such rhizobial species that nodulates the stems of *Sesbania rostrata*. Rhizobium cells contain genes for nitrogen fixation (nif genes) on a megaplasmid. The bacteria enter the roots through root hairs; interaction is progressing through several steps and it ultimately leads to nodule formation. Inside the nodule many bacterial cells changing into nondividing bacteroids, which produces nitrogenase enzyme which reduces atmospheric nitrogen to ammonia.

Uses

- Rhizobium can fix 50-200 kgs N/ha in one crop season.
- It can increase yield up to 10-35%.
- Due to rhizobial activities, the root hairs and nodules secrete a mucous substance which enhances the soil fertility and growth of the plant.
- The enzyme nitrogenase will reduce the molecular nitrogen to ammonia which is readily utilized by the plant.

- By means of seed treatment, the germination of seeds gets stimulated and in turn increase crops yield potential.

Methods to use

- For 10 kg of seed 1 pocket of rhizobium (200 g) is sufficient.
- Mix this rhizobium in well fertile soil along with 200 ml rice kanchi and mix well
- Shade dry for 30 minutes and take up sowing immediately

Azospirillum

It is the associative symbiotic nitrogen fixer, aerobic free living which makes the atmospheric nitrogen available to various crops. This nitrogen-fixing bacterium when applied to the soil undergoes multiplication in billions and fixes atmospheric nitrogen in the soil. Nitrogen fixation in the rhizosphere through the action of nitrogenase enzyme. The *Azospirillum* form associative symbiosis with many plants particularly with those having the C4-dicarboxylic pathway of photosynthesis (Hatch and Slack pathway), because they grow and fix nitrogen on salts of organic acids such as malic, aspartic acid. Thus it is mainly recommended for maize, sugarcane, sorghum, pearl millet etc. The Azotobacter colonizing the roots not only remains on the root surface but also a sizable proportion of them penetrates into the root tissues and lives in harmony with the plants. They do not, however, produce any visible nodules or out growth on root tissue.

Uses

- *Azospirillum* sp. have the ability to fix 20-40 Kgs N/ha.
- It results in average increase in yield of 15-30%.
- It also results in increased mineral and water uptake.
- It also promotes root development and vegetative growth.
- Fortification of the soils occurs with bacterial metabolites and by secreting growth promoters.

It is recommended for Paddy, Millets, Oilseeds, Fruits, Vegetables, Sugarcane, Banana, Coconut, Oil palm, Cotton, Chilly, Lime, Coffee, Tea, Rubber, Flower, Spices, Herbs, Ornaments, trees etc.

Methods to use

Seed treatment

For 1 acre seeds, 2 pockets. Mix well with rice kanji and shade dry for 30 minutes.

For transplanted crop

4 pockets of *Azospirillum* should be mixed with well decomposed FYM and broadcast it over 1 acre

Seedling root tip

Two packets of the inoculant are mixed in 40 litres of water. The root portion of the seedlings required for an acre is dipped in the mixture for 20 minutes and then transplanted.

For trees

For a grown tree 20 to 50 of *Azospirillum* should be mixed with well decomposed farm yard manure and apply over the root zone.

Azotobacter

It belongs to family ***Azotobacteraceae***, aerobic, free living, and heterotrophic in nature. *Azotobacte*r are present in neutral or alkaline soils and *A. chroococcum* is the most commonly occurring species in arable soils. *A. vinelandii, A. beijerinckii, A. insignis* and *A. macrocytogenes* are other reported species. The number of *Azotobacter* rarely exceeds of 104 to 105 g-1 of soil due to lack of organic matter and presence of antagonistic microorganisms in soil. The bacterium produces anti-fungal antibiotics which inhibits the growth of several pathogenic fungi in the root region thereby preventing seedling mortality to a certain extent. The population of *Azotobacter* is generally low in the rhizosphere of the crop plants and in uncultivated soils. The occurrence of this organism has been reported from the rhizosphere of a number of crop plants such as rice, maize, sugarcane, bajra, vegetables and plantation crops.

Blue Green Algae

Blue Green Algae (Cyanobacteria) are photosynthetic, free living and prokaryotic organisms which fix nitrogen a symbiotically but some cyanobacteria are known to form symbiotic associations. Examples of cyanobacteria are Anabaena, Nostoc, Plectonema etc. Cyanobacteria produce nitrogenase and nitrogen fixation occurs in specialized structures called heterocysts. Also heterocysts act as oxygen proof compartments which protect nitrogenase from oxygen inactivation. In India the soil conditions in rice fields provide suitable environment for growth of cyanobacteria. Cyanobacterial biofertilizers can be prepared by farmers in their own field. For this open tanks of galvanized iron sheets or bricks and cement are made and cyanobacteria of required strain are cultured in this tank. A mixture of sieved soil, super phosphate, sodium molybdate and water is added into tank then mixture is thoroughly mixed and allowed to stand for 24 hrs. Then starter culture of required strain of cyanobacteria is sprinkled on surface of water and allowed to stand for 15-20 days. Then thick serum of algal mass is collected and allowed to dry.

Uses

- Cyanobacteria add growth promoting substances including Vitamin B 12 and it also improves soil aeration and water holding capacity.
- It can provide 20-30 kg of biologically fixed nitrogen per hectare under normal conditions which can be doubled if optimum amounts of phosphate and molybdenum are available in soil.
- It provides partial tolerance to pesticides and fungicides.
- BGA is also supplied with Azolla (an aquatic fern), which harbors Anabaena azolle (a BGA) in leaf cavities, providing symbiotic association.

Method of application

BGA are added to the soil within ten days of transplanting at the rate of 10 -12 kg/ha. They are available as small bits in plastic packets. This should be powdered and directly added to the soil. Water should be allowed to stagnate to a depth of 3–5 cms in fields where algae are grown. Blue green algae should be added to the field continuously for four cropping seasons. Thereafter, it grows naturally in the soil and produces the desired results.

Azolla

It is an aquatic heterosporus fern which contains an endophytic cyanobacterium Anabaena azollae in its leaf cavity. These belongs to eight different families, phototrophic in nature and produce Auxin, Indole acetic acid and Gibberllic acid, fix 20-30 kg N/ha in submerged rice fields as they are abundant in paddy, so also referred as "paddy organisms". N is the key input required in large quantities for low land rice production. Soil N and BNF by associated organisms are major sources of N for low land rice4. The 50-60% N requirement is met through the combination of mineralization of soil organic N and BNF by free living and rice plant associated bacteria. To achieve food security through sustainable agriculture, the requirement for fixed nitrogen must be increasingly met by BNF rather than by industrial nitrogen fixation.

BGA forms symbiotic association capable of fixing nitrogen with fungi, liverworts, ferns and flowering plants, but the most common symbiotic association has been found between a free floating aquatic fern, the Azolla and *Anabaena azollae* (BGA). *Azolla* contains 4-5% N on dry basis and 0.2-0.4% on wet basis and can be the potential source of organic manure and nitrogen in rice production. The important factor in using *Azolla* as biofertilizer for rice crop is its quick decomposition in the soil and efficient availability of its nitrogen to rice plants. Besides N-fixation, these biofertilizers or biomanures also contribute significant amounts of P, K, S, Zn, Fe, Mb and other micronutrient. The fern forms a green mat over water with a branched stem, deeply bilobed leaves and roots. The dorsal fleshy lobe of the leaf contains the algal symbiont within the central cavity. Azolla can be applied as green manure by incorporating in the fields prior to rice planting. The most common species occurring in India is *A. pinnata* and same can be propagated on commercial scale by vegetative means. It may yield on average about 1.5 kg per square meter in a week. India has recently introduced some species of Azolla for their large biomass production, which are *A.caroliniana, A. microphylla, A.filiculoides and A. mexicana.*

Uses

- It is mostly used in rice fields where water is available for its growth and multiplication.
- It is supplemented with 8-20 kg phosphate per hectare. It improves the height of rice plants, number of tillers, grains and straw yield.
- There is a 50% higher yield by using *Azolla* as biofertilizer.

Method of application

Growing algae along with paddy supplies the crop with the required nitrogen and phosphorus. Azolla or blue green algae should be strewn in the field 5–10 days after transplantation of paddy. The field should be drained twenty-five days and 45–50 days after strewing and the algae should be stamped into the soil.

Five to seven kilos of azolla are required per hectare of land. First, the water should be allowed to stagnate in the land that is to be cultivated. Then azolla should be sown. After one week, these plants are stamped into the soil before transplanting is done. It degrades in about 7–10 days and thereby provides nitrogen to the rice crop.

B. Phosphate solubilizers

Different bacterial species solubilize insoluble inorganic phosphate compounds, such as tricalcium phosphate, dicalcium phosphate, hydroxyapatite, and rock phosphate. Among the bacterial genera with this capacity are *pseudomonas, Bacillus, Rhizobium, Burkholderia, Achromobacter, Agrobacterium, Micrococcus, Aerebacter, Flavobacterium* and *Erwinia.* There are considerable populations of phosphate solubilizing bacteria in soil and in plant rhizospheres. These include both aerobic and anaerobic strains, with a prevalence of aerobic strains in submerged soils. A considerably higher concentration of phosphate solubilizing bacteria is commonly found in the rhizosphere in comparison with non rhizosphere soil. The soil bacteria belonging to the genera *Pseudomonas* and *Bacillus* and *Fungi* are more common

C. P Mobilizing Biofertilizers

The term Mycorrhiza denotes "fungus roots". It is a symbiotic association between host plants and certain group of fungi at the root system, in which the fungal partner is benefited by obtaining its carbon requirements from the photosynthates of the host and the host in turn is benefited by obtaining the much needed nutrients especially phosphorus, calcium, copper, zinc etc., which are otherwise inaccessible to it, with the help of the fine absorbing hyphae of the fungus. These fungi are associated with majority of agricultural crops, except with those crops/plants belonging to families of *Chenopodiaceae, Amaranthaceae, Caryophyllaceae, Polygonaceae, Brassicaceae, Commelinaceae, Juncaceae and Cyperaceae.*

Mycorrhizal fungi

The term Mycorhiza was coined by Frank in 1885. It is a distinct morphological structure which develops as a result of mutualistic symbiosis. It is the symbiotic association between plant roots and soil fungus. The kinds of mycorrhizae are:

1. **Ectomycorriza:** Ectomycorrhiza are found on the roots of forest trees (e.g. pine, oak, beech, eucalyptus, etc.). In general root hairs are absent on roots of some higher plants. Therefore roots are infected by mycorrhizal fungi and form a mantle. The hyphae grow intercellularly and develop Hartig net in cortex. So a bridge is established between the soil and root through the mycelia. They absorb nitrogen, phosphorus, potassium and calcium and produce growth promoting substances i.e. cytokinins. The major functions which ectomycorrhizae perform are: absorption of water; solubilisation of complex organic molecules into simplex inorganic nutrients then their absorption and transfer to roots; protection of plants from attack of inciting pathogens by secreting antimicrobial substances.
2. **Endomycorrhiza:** In this association fungus does not form an external sheath or mantle.

   The fungus lives in the intercellular spaces as well as intracellularly in the cortical cells of roots. Only a small portion of fungus lives outside the root. Endomycorrhiza are found in the roots of most fruits and other horticultural crops e.g. coffee, pepper, cardamom, and betelvine. They particularly help in phosphorus nutrition. They also produce growth promoting substances and offer resistance against pathogens.

3. **Vesicular Arbuscular Mycorrhiza:** One of the important type of endomycorrhizae, VAM plays a great role in inducing plant growth. VAM is highly versatile and colonizes 85% of the plant families. It penetrates the roots, forms arbuscules and vesicles in the cortical cells of the roots Vesicles are thick walled swollen structures and arbuscules are branched hautorial branches of mycelium .They serve as food storage organs of the fungus. VAM can be produced on a large scale by pot culture technique.

Uses

- VAM fungi enhances water uptake in plants.
- They increase resistance in plants and reduce the effects of pathogens and pests on plants.
- The mychorrhiza penetrates the roots, mobilizes & supplies phosphorus and other micronutrients to the plants.
- VAM fungi reduce plant response to soil stress.
- Mycorrhizal plants show higher tolerance to high soil temperatures and heavy metal toxicity.

Method to use

AM Inoculation

Optimal spore count 60-100 spores/ 100g soil

Rate of Inoculation

Vegetables	-	100 g/m^2 nursery
Fruit trees & Coconut	-	100-200 g/tree
Other crops	-	10% of the seed rate
Established plants	-	10 g/plant
Nursery	-	750-1000 g /m^2
Nursery poly bag	-	10 g/ poly bag

C. Zinc solubilizers

The nitrogen fixers like *Rhizobium, Azospirillum, Azotobacter, BGA* and Phosphate solubilizing bacteria like *B. megaterium, Pseudomonas striata,* and phosphate mobilizing Mycorrhiza have been widely accepted as bio-fertilizers. However these supply only major nutrients but a host of microorganism that can transform micronutrients are there in soil that can be used as bio-fertilizers to supply micronutrients like zinc, iron, copper etc.,The zinc can be solubilized by microorganisms viz., *B. subtilis, Thiobacillus thioxidans and Saccharomyces* sp. These microorganisms can be used as bio-fertilizers for solubilization of fixed micronutrients like zinc. The results have shown that a *Bacillus* sp. (Zn solubilizing bacteria) can be used as bio-fertilizer for zinc or in soils where native zinc is higher or in conjunction with insoluble cheaper zinc compounds like zinc oxide (ZnO), zinc carbonate (Z_nCO_3) and zinc sulphide (ZnS) instead of costly zinc sulphate.

D. Silicate solubilizing bacteria (SSB)

Microorganisms are capable of degrading silicates and aluminum silicates. During the metabolism of microbes several organic acids are produced and these have a dual role in silicate weathering. They supply H+ ions to the medium and promote hydrolysis and the organic acids like citric, oxalic acid, Keto acids and hydroxy carboxylic acids which from complexes with cations, promote their removal and retention in the medium in a dissolved state.

The studies conducted with a Bacillus sp. isolated from the soil of granite crusher yard showed that the bacterium is capable of dissolving several silicate minerals under *in vitro* condition. The examination of anthrpogenic materials like cement, agro inputs like super phosphate and rock phosphate exhibited silicate solubilizing bacteria to a varying degree. The bacterial isolates made from different locations had varying degree of silicate solubilizing potential. Soil inoculation studies with selected isolate with red soil, clay soil, sand and hilly soil showed that the organisms multiplied in all types of soil and released more of silica and the available silica increased in soil and water. Rice responded well to application of organic siliceous residue like rice straw, rice husk and black ash @ 5 t/ha. Combining SSB with these residues further resulted in increased plant growth and grain yield. This enhancement is due to increased dissolution of silica and nutrients from the soil.

E. Potash mobilizers

Frateuria aurantia is known as Potash Mobilising Bacteria (PMB) belonging to the family Pseudomondaceae. Potash Mobilizing Bacteria has great roll as for plant growth it is usually abundant in soil. Total Potash Mobilizing Bacteria contents in soil range between 3000 to 100,000 kg/ha in the upper 0.2 m of the soil layer.

Potash Mobilizing Bacteria plays a vital role in the formation of amino acids and proteins from ammonium ions, which are absorbed by roots, from the soil. it is also responsible for the transfer of carbohydrates, proteins, etc. from the level to the roots. It also plays a vital role in the uptakes of other elements particularly nitrogen, phosphorus and calcium, Potash Mobilizing Bacteria regulates the permeability of the cellular membrane. It activates number of enzymes, e.g. alcohol dehydrogenase and its deficiency decreases photosynthesis. Potash Mobilizing Bacteria increases the resistance of crops to hot and dry conditions and insect pest and diseases. It increases the stiffness of straw in cereals and therefore the loading of cereal is reduced. It improves the quality of fruits and grains.

Potash mobilizers at the rate of 200 ml/ acre could be applied for all crops.

Table 7. A key to biofertilizer use

For Crops	Biofertilizers recommended & Doses
Pulse crops like moong, urad, arhar, cowpea, lentil, pea, bengal gram, all beans, ground nut, soybean, leucern, berseem and other legume crops.	*Rhizobium* 200 gm + PSB 200 gm for every 10 kg of seed as seed treatment.
All non-legume crops like wheat, seedsown upland paddy, barley, maize, cotton,sorghum, bhindi,	*Azotobacter* 200 gm + PSB 200 gm for every 10 kg of seed as seed treatment.

(Contd.)

mustard, sunflower, niger etc. and other non legume crops taken by direct seed sowing.	
Jute	*Azospirillum* 200 gm + PSB 200 gm for every 10 kg of seed as seed treatment.
Vegetables like tomato, brinjal, chilli, cauliflower, cabbage etc. and othertransplanted crops.	*Azotobacter* 1kg + PSB 1 kg for one acre as seeding root dip method.
Lowland transplanted paddy.	*Azospirillum* 2 kg + PSB 2 kg for one acre as seedling root dip for 8 -12 hours.
Potato, ginger, colocasia, turmericand jhum paddy.	*Azotobacter* or *Azospirillum* 4 kg + PSB 4 kg/acre mixed with 100-200 kg compost and applied in soil.
Standing plantation crops like tea, coffee, rubber, mulberry and fruit trees.	2 - 3 kg *Azotobacter/Azospirillum* + 2-3 kg PSB mixed with 200 kg Compost for one acre and applied as soil treatment. This treatment is to be done 2 to 3 times a year with a gap of 4- 6 months
Sugarcane	5 kg *Acetobacter* mixed in sufficient water for setts dipping treatment

Advantages of Using Biofertilizers in Agriculture

- It is a low cost and easy technique.
- The biofertilizers increase 15-35% additional yield in most of vegetable crops.
- Besides fixing atmospheric nitrogen, cyanobacteria synthesize and Secrete several growth hormones (auxins and ascorbic acid) and vitamins which enhance seed germination and growth of crop plants.
- They do not cause atmospheric pollution and increase soil fertility
- Some biofertilizers excrete antibiotics and thus act as pesticides.
- They improve physical and chemical properties of soil such as water holding capacity, buffer capacity etc.,

- Some of the biofertilizers enhance crop yield even under ill irrigated conditions where
- chemical fertilizers are of not much advantage.
- They are eco friendly and pose no danger to the environment.

Multi varietal seed technique (Navathaniyam)

Soil fertility can be enhanced non- chemically using multi varietal seed technique with in 6.5 month (200 days) even under highly chemical fertilizer applied and deteriorated soil conditions.

Mixed sowing of the following five crops is called as multi varietal seed technique

1. Cereals: Any four crops under this category

Example: Sorghum (*Sorghum bicolor*) - 1 kg

Pearl millet (Cumbu - *Pennisetum glaucum*) - ½ kg

Foxtail millet (Thenai - *Setaria italica*) - ¼ kg

Little millet (Samai - *Panicum sumatrense*)- ¼ kg

2. Pulses: Any four crops under this category

Example: Blackgram (*Vigna mungo*) -1 kg

Green gram (*Vigna radiata*) - 1 kg,

Cowpea (*Vigna unguiculata*) - 1 kg,

Bengal gram (*Cicer arietinum*) - 1 kg

3. Oilseeds: Any four crops under this category

Example: Gingelly (*Sesamum indicum*) - ½ kg

Groundnut (*Arachis hypogaea*) - 2 kg

Sunflower (*Helianthus annuus*) - 2 kg,

Castor (*Ricinus communis*) - 2 kg

4. Green manures: Any four crops under this category

Eg: Dhaincha (*Sesbania aculeata*) - 2 kg

Sun hemp (*Crotalaria juncea*) - 2 kg

Wild indigo(Nari payaru - *Baptisia australis*) - ½ kg

Horse gram (*Macrotyloma uniflorum*) - 1 kg

5. Spices and condiments: Any four crops under this group

Eg: Mustard (*Brassica juncea*) - ½ kg

Fenugreek (*Trigonella foenum-graecum*) - ¼ kg

Cumin (*Cuminum cyminum*) - ¼ kg

Coriander (*Coriandrum sativum*) - 1 kg

Grow the above crop combinations as sole pure crop upto 60 days and incorporate into the soil through substantial ploughing which provides all major and micro nutrients to the soil and support soil life and this in turn better crop growth. This technique rejuvenuate the soil through improvement its physical, chemical and biological properties, enhances biomass addition to the soil and helps to restore its fertility.

Green manuring

Green manuring is a very ancient farm practice. Green manures have been used in traditional agriculture for thousands of years but conventional farming systems largely rejected them as the use of fertilisers and pesticides became more common. Although they have many roles they are still often under utilised by today's organic farmers. Modern agriculture techniques are inorganic promotive which are more expensive. So there is an urgent need to identify the alternate nitrogen sources to supplement inorganic fertilizers. Green manuring is a low cost but effective technology in increasing the productive capacity of the soil.

Benefits of using green manuring

Green manuring offers an inexpensive way of improving crop yields and it takes little extra effort. Green manures are especially important on farms where there is not enough animal manure available, and when it is not possible to bring in natural fertilisers from elsewhere. Although the use of green manures may seem to create extra work, they do provide a number of benefits:

Erosion

High rainfall and substantial slopes on agricultural land create ideal conditions for erosion to occur. Rain drops hitting bare ground cause soil particles to be dislodged. These are transported downhill in water flow, and pick up more soil as momentum builds. But covered soil is less vulnerable to water erosion. Wind erosion is reduced as the green manure increases surface roughness reducing the wind speed close to the soil. The root system also has a binding effect on the soil. The green manure also reduces run off substantially, at ground covers of greater than 75%, reducing erosion by rain.

Deposition of Sediment

Increase of cover reduces upland erosion, which in turn reduces sediment from floodwaters and wind.

Compaction

Increased biomass, when decomposed, increases organic matter promoting increased microbial activity and aggregation of soil particles. This increases soil porosity and reduces bulk density. Caution: plant cover crops when soils are not wet, or use other methods such as aerial seeding. Some species also produce deep tap roots which help break up compacted soil

Soil aggregation at the surface

Aggregate stability will increase with the addition of and the decomposition of organic material by microorganisms. Green manures can improve soil structure in a number of ways. The extensive fine roots of some, such as rye, enmesh the soil, helping to stabilise aggregates and increasing pore size thus improving seedbed structure. A key function of green manures is the addition of organic matter to the soil. They do this whilst still growing, producing root exudates which provide food for microorganisms, which in turn produce polysaccharide gums, which "glue" soil aggregates together. They may also provide a bridge between mycorrhizal crops in order to maintain a high population of soil mycorrhiza, which help maintain soil structure, again by enmeshing soil aggregates. Brassicas and lupins, however, are non mycorrhizal and will break that bridge.

Some green manures species are particularly good at improving problems with soil structure. Lucerne, chicory and sweet clover have a strong tap root that can break through compacted soils or those with a plough pan. The fibrous root system of cocksfoot is good at improving soil crumb structure.

Infiltration

Surface cover reduces erosion and runoff. Cover crop root channels and animal activities, such as earthworms, form macropores that increase aggregate stability and improve infiltration. Caution: Macropores can result in an increase in leaching of highly soluble pesticides if a heavy rain occurs immediately after application. However, if only sufficient rainfall occurs to move the pesticide into the surface soil after application, the risks for preferential flow are minimal. Cover crops, especially small grains, utilize excess nitrogen.

Improved soil moisture

The soil cover, or mulch, that is provided by a green manure also greatly improves drought resistance. The green manure residues add organic matter to the soil, which increases infiltration of water into the soil and increases the water-holding capacity of the soil. Green manures improve soil structure, letting more air into the soil and improving drainage.

Soil Crusting

Cover crops will provide cover prior to planting the main crop. If conservation tillage is used, benefits will continue after planting of main crop. Increases of organic matter, improved infiltration, and increased aggregate stability reduce soil crusting.

Nutrient Loss or Imbalance

Decomposition of increased biomass provides a slow release of nutrients to the root zone. Legume cover crops fix atmospheric nitrogen and provide nitrogen for the main crop. Legumes utilize a higher amount of phosphorus than grass or small grains. This is useful in animal waste utilization and management. Small grains are useful as catch crops to utilize excess nitrogen, which reduces the potential for nitrogen leaching. Caution: To prevent nutrient tie ups, cover crops should be killed 2-3 weeks prior to planting main crop.

Adding organic matter to the soil

Cover crops reduce run-off resulting in reduced nutrient and pesticide losses from surface runoff and erosion. Increased organic matter improves the environment for soil biological activity that will increase the breakdown of pesticides.

Organic Matter

Decomposition of increased biomass results in more organic matter. Research shows cover crops killed 2-3 weeks prior to planting main crop, results in adequate biomass and reduces the risk of crop losses from soil moisture depletion and tie up of nutrients. Green manure that have a high potential for biomass production allow soil organic matter levels to be maintained and/or increased in a way that is both practical and economical.

Biological Activity

Cover and green manure crops increase the available food supply for microorganisms resulting in increased biological activity.

Ameliorating problem soils

- *Sesbania aculeata* (dhaincha) applied to sodic soils continuously for four or five seasons improves the permeability and helps to reclaim.

- *Argemone mexicana* & *Tamarindus indica* has a buffering effect when applied to sodic soils.

Fertility improvement of soils

Legumes (plants that produce their seeds inside pods) are able to fix nitrogen (N) from the atmosphere into a plant-usable form that accumulates in plant tissues. Legumes can thereby add large quantities of nitrogen to farmers' soils. Most of the widely used legume green manures are capable of producing more than 50 kg N/ha, while a few green manure species fix significantly more. 4. Increase the solubility of lime phosphates, trace elements etc., through the activity of the soil microorganisms and by producing organic acids during decomposition.

Improvement in crop yield and quality

- Increases the yield of crops to 15 to 20 %
- Vitamin and protein content of rice increased

Improvement of soil tilth

- Green manuring builds up soil structure and improves tilth
- Promotes formation of crumbs in heavy soils leading to aeration and drainage
- Increases the water holding capacity of light soils.
- Form a canopy cover over the soil, reduce the soil temperature and prevents from erosive action of rain and water

Disease control

It is well established that green manures can suppress disease although these effects are variable and the mechanisms not fully understood. The way in which green manures suppress disease fall broadly into two categories: green manures can either provide organic matter to sustain microbial communities that suppress pathogens or the green manure may have a direct biocidal effect on the pathogen. Green manures may support microbial communities of bacteria, non-pathogenic Fusarium species, streptomyces and other actinomycetes. These can suppress pathogens through competition, antibiosis, parasitism or by inducing systemic resistance in plants.

Although green manures can act as a disease suppressive agent, care must be taken that they do not also act as a green bridge for diseases. The most obvious danger of this is growing a brassica green manure such as mustard in a rotation that already contains brassicas, as this will greatly increase the risk of introducing persistent soilborne diseases such as clubroot

Addition of Nitrogen

Nitrogen is added to the soil by nitrogen fixation if the green manure crop is a legume. Nitrogen fixation is a process by which nitrogen in the air is converted into ammonium nitrogen. Legumes are members of the family Leguminosae (also named Fabaceae) and have the capacity in symbiosis with certain soil microorganisms to convert gaseous nitrogen from the air into nitrogenous chemical compounds derived from the ammonium that is fixed. The nitrogen is fixed in nodules in the roots and is used by the legume for root growth and shoot growth. Most of the fixed nitrogen, about two-thirds, however, ends up in the shoots. If the shoots are incorporated, decomposition of the organic matter releases the fixed nitrogen by mineralization. The amount of nitrogen added varies with the kind of legume, depending largely on the amount of shoot dry matter that is produced by the crop.

When initiating the organic farming, to obtain the actual benefit of nitrogen from green manure, it may be necessary to let the legume grow on the land for at least one year; otherwise, the fixation of nitrogen will be much lower than that noted, perhaps half or less.

Types of green manures

Legumes

These are generally considered to be nitrogen fixing but this will only happen in the presence of correct strains of Rhizobium bacteria. For more common legume species these will certainly be present naturally in the soil but some more unusual species may benefit from inoculating the seed before sowing.

Non Legumes

These will not fix nitrogen but can be very effective at preventing nitrate leaching, adding organic matter to the soil and smothering weeds.

Selecting which green manures to use

There are many plants which can be used as green manures. In particular, the type of green manure should be selected according to the type of crop it is growing with or in between. For a large plant like maize, a large green manure like velvet bean or Sesbania should be used. For a short crop like many vegetables, smaller green manures such as mustard or buckwheat can be used.

- The relative cost of seed
- Speed of germination and growth
- Longevity
- Multipurpose use
- Tolerance to mowing or grazing
- Suitability for animal forage
- Competitiveness with weeds
- Ability to grow in a nutrient-poor soil
- Nitrogen fixing ability
- Pest/disease resistance
- Ease of incorporation
- Speed of decomposition
- Unlikeliness to return as a weed in the next crop
- Short duration, fast growing, high nutrient accumulation ability
- Tolerance to shade; flood, drought and adverse temperatures.
- Wide ecological adaptability
- Efficiency in use of water
- Early onset of biological nitrogen fixation
- High N accumulation rates

- Timely release of nutrients
- Ability to cross-inoculate or responsive to inoculation
- High N sink in underground plant parts.
- A green manure must suit the local climate, and the soil that it is to be sown in. This will help to keep the green manure healthy and to keep pests and diseases to a minimum.

Green manures

Cowpea (*Vigna unguiculata*)

Cowpea is one of the important leguminous green leaf manure crops. As this plant is easily decomposable and very well suited for green manure purpose. June-July months are best suited for sowing of this manure. Even though it is being cultivated in summer months (March to April). Use of effective *Rhizobium* bacteria increase the fixation of nitrogen up to 40 kg/ha.

Dhaincha (*Sesbania aculeata*)

Dhaincha is suitable for loamy and clayey soils. It is fairly resistant to drought as well as stagnation of water. It grows well even in alkaline soils and corrects alkalinity if grown repeatedly for 4-5 years. The roots have plenty of nodules. It yields about 10-15 tonnes of green manure per ha and requires a seed rate of 30-40 kg/ha. Use of effective Rhizobium strain with seeds fixes the Nitrogen 1 kg / day.

Sesbania speciosa

It is a valuable green manure for wetlands and can be grown in a wide range of soils. Seed production is prolific however, pods are frequently attacked by insects. This green manure can be raised on the field borders. Sesbania seedling (21days) can be planted in a single line at 5-10 cm apart in the borders of the fields. In about 90 days it produces about 2-4 tonnes of green manure per ha. It does not affect the rice yield by shading or root effect. If second rice crop is planted immediately after the first crop, the manure can be incorporated into the field. About 300-400g of seeds are sufficient to raise nursery and plant the seedlings around the boundary of one hectare. To control insects *Verticillium lecanii* (Liquid) fungi is useful.

Sunnhemp (*Crotalaria juncea*)

It is a quick growing green manure crop and gets ready for incorporation in about 45 days after sowing. It does not withstand heavy irrigation leading to flooding. The crop is at times subject to complete damage by leaf eating caterpillars. The crop can produce about 8-12 tonnes of green biomass per ha. The seed requirement is 30 kg/ha.

Manila agathi (*Sesbania rostrata*)

One of the important features of this green manureis that in addition to the root nodules, it produces nodules in the stem. The stem nodulation is an adaptation for waterlogged situation since flooding limits growth of green manures and may reduce root nodulation. Under normal condition, both root and stem nodules are effective in N fixation. It has higher N content of 3.56% on dry weight basis. Biomass production is higher during summer (April – June) than in winter (Dec. – Jan.) season. This green manure can also be produced by raising seedlings (30 days old) and planted in the paddy field along the bunds or as intercrop with rice. Use of *Rhizobium* bacteria increase the nitrogen fixation about 60-100 kg/ha/year.

Wild indigo (*Tephrosia purpurea*)

This is a slow growing green manure crop and cattle do not prefer to graze it . The green manure is suitable for light textured soils, particularly in single crop wetlands. It establishes itself as a self sown crop and the seeds remain viable till the harvest of rice. On an average about 3-4 tonnes of green manure is obtained in one ha. The seed rate is 30 kg/ha. The seeds have a waxy impermeable seed coat and hence scarification is required to induce germination. Soaking seeds in boiling water for 2-3 minutes is also equally effective in promoting germination.

Indigo (*Indigofera tinctoria*)

It resembles wild indigo and is along duration crop with more leafy growth. It comes up well in clayey soils with one or two irrigations.

Pillipesara (*Phaseolus trilobus*)

This is a dual purpose crop yielding good fodder for the cattle and green manure. Pillipesara comes up well in hot season with sufficient soil moisture. Loamy or clayey soils are best suited. After taking one or two cuttings for fodder or light grazing by animals, the crop can be incorporated into the soil. About 5-8 tonnes of manure can be obtained from one ha.

Gliricidia (*Gliricidia maculata*)

This is a shrubby plant thatcomes up well in moist situations. Under favourable conditions, it grows well like a tree. It can be easily grown in waste lands, farm road sides, field bunds, etc. The crop can be established by stem cuttings or seedlings planted in the field borders. It can be pruned for its tender loppings and compound leaves for green leaf manuring at the time of puddling rice. On an average, a well-established plant yields 12-15 kg green matter. About 400 plants on the peripheral bunds yields 5-6 tonnes green manure/ha.

Karanj (*Pongamia glabra*)

It is a leguminous tree grown in wastelands. On an average, a tree can yield 100-120kg of green matter. The leaves contain about 3.7% N (on dry weight basis).

Calotropis (*Calotropis gigantca*)

On roadsides and fallow lands, the plant grows wild under different soil and climatic conditions. The leaves are more succulent and a plant can produce about 4-5 kg of green matter. Besides it also helps in controlling soil borne pests like termite.

Panchagavya – An effective on farm input

Panchagavya is a bio promoter with a combination of five products obtained from the cow, which includes cow dung, cow's urine, milk, curd and ghee. All the five products are individually called gavya and collectively termed as *Panchagavya*. *Panchagavya* has the potential to play the role of promoting growth and providing immunity in plant system. The *Panchagavya* acts as growth promoter (75%)

and immunity booster (25%) and exactly fills the missing link to sustain the organic farming without any yield loss. It plays an important role in providing resistance to diseases, pests and in increasing the overall yield due to the immunostimulant activity of the cow dung and it can be prepared by the farmers themselves with the materials available on the farm.

Panchagavya

• Fresh cow dung	5 Kg
• Cow Urine	3 lit
• Cow milk	2 lit
• Curd	2 lit
• Cow desi ghee	1 kg

Method of preparation

Mix all the ingredients thoroughly and ferment for 15 days with twice stirring per day. Dilute 3 lit of Panchagavya in 10 lit water and spray over soil. 20 lit *Panchagavya* is needed per acre for soil application along with irrigation water. Panchagavya can also be used for seed treatment. Soak seeds for 20 minutes before sowing.

Fig.8. Products used in making of Panchagavya

Table 8. Panchagavya spray schedule for various crops

Crops		Time schedule
Rice	:	10,15,30 and 50th days after transplanting
Sunflower	:	30,45 and 60 days after sowing
Blackgram	:	Rainfed : 1st flowering and 15 days after flowering
	:	Irrigated : 15, 25 and 40 days after sowing
Green gram	:	15,25,30,40 and 50 days after sowing
Castor	:	30 and 45 days after sowing
Groundnut	:	25 and 30th days after sowing
Bhendi	:	30,45,60 and 75 days after sowing
Moringa	:	Before flowering and during pod formation
Tomato	:	Nursery and 40 days after transplanting, seed treatment with 1% for 12 hrs
Onion	:	30,45 & 40 days after transplanting
Rose	:	At the time of pruning & budding
Jasmine	:	Bud initiation and setting
Vanilla	:	Dipping sets before planting
Fruit trees		
First time	:	One month before flowering
Second time	:	15 days after flowering
Third time	:	When the fruit is in peas size
Fourth time	:	After harvest once
Leafy crops		
Greens, curry leaf, tea, coffee,	:	2 % weekly once

Modified panchagavya

Panchagavya, by name constitutes five products evolving from cow. Rural community Action Centre (NGO) has modified this Panchagavya by adding a few more ingredients and this modified version has a lot of beneficial effects on a variety of crops and livestock. The present form of Panchagavya standardized by TNAU is a single organic input which can act as a potentiator. It is essentially a product containing the following:

Table 9. Ingredients of Modified panchagavya

Sl.No.	Ingredients	RCAC	TNAU Standard
1.	Cow dung slurry (from Gobar gas plant)	4 Kg	-
2.	Fresh cow dung	1 Kg	5.0 kg
3.	Cow's urine	3 litre	3.0 lit
4.	Cow's milk	2 litre	2.0 lit
5.	Cow's curd	2 litre	2.0 lit
6.	Cow's ghee	1 Kg	1.0 kg
7.	Sugarcane juice	3 litre	3.0 lit
8.	Tender coconut water	3 litre	3.0 lit
9.	Banana (ripe)	12 Nos	1.0 kg

Method of preparation

First day

Fresh cow dung	-5 kg
Cow ghee	- ½ kg

Mix well these two and keep under shade for three days. Stir it well twice a day

Fourth day

Cows milk	- 2 litre (boiled and cooled)
Curd	- 2 litres
Cow's urine	- 3 litres
Tender coconut	3 litres
Jaggery	- ½ kg (dissolved in 3 litres of water)
or Sugarcane juice	- 3 litres
Banana	- 12 nos

Ferment the Tender coconut for one week and take 2 litres

On forth day, mix all the ingredients thoroughly and ferment for 15 days with twice stirring per day in a wide mouthed mud pot, concrete tank or plastic cans. The container should be kept open under shade condition and the contents of the container. Dilute 3 lit of Panchagavya in 100 lit water and spray over soil. 20 lit *Panchagavya* is needed per acre for soil application along with irrigation water. Panchagavya can also be used for seed treatment. Soak seeds for 20 min before sowing.

The products of local breed of cow is said to have more potency than exotic breeds and care should be taken not to mix buffalo products.

Methods of application

The *Panchagavya* can be applied in different ways like foliar spray through irrigation and as seed / seedling treatment. About 3% solution has been found to be the most effective in increasing the production and consumption of organic produces.

As foliar spray

Investigations prove that 3% solution was found to be the most effective compared to the higher and lower concentration. Three litres of panchagavya to every 100 litres of water is the ideal recommended dosage for all the crops. The power sprayer of 10 litres capacity may need 300 ml/tank. When sprayed with power sprayer, sediments are to be filtered and when sprayed with hand operated sprayers, the nozzle with higher pore size has to be selected.

Through Flow / irrigation system

In the case of flow or irrigation systems the solution of *Panchagavya* can be mixed with irrigation water at 50 litres per hectare either through drip or flow irrigation water at 50 litres per hectare either through drip or flow irrigation.

As Seed / seedling treatment

3% solution of *Panchagavya* can be used to treat the seeds and seedlings by soaking and dipping before planting. Soaking for 20 minutes is sufficient. Rhizomes of turmeric, ginger and sets of sugarcane can be soaked for 30 minutes before planting. In the case of seed storage 3% panchagavya solution can be used to dip the seeds before drying and storing them.

For seed storage

Panchagavya solution of 1 % can be used to dip the seeds before drying and storing them.

Table 10.Frequency of spray

Pre flower phase (20 days after planting)	Once in 15 days, two sprays depending upon duration of the crops.
Flowering and pod setting stage	Once in 10 days, two sprays
Fruit / pod maturation stage	Once during pod maturation

Enriched Panchagavya (for one acre quantity required 200 kg)

i)	Dung (cow/buffalo/ox)	: 40 kg
ii)	Urine (cow/buffalo/ox)	: 40 kg
iii)	Butter/mustard oil	: ½ litres
iv)	Milk	:2 ½ litres
v)	Lassi/diluted curd	: 8 Ltrs.
vi)	Gram flour and Methi flour	: One kg each
vii)	Old Jaggery (gur)	: 2 kg
viii)	Ripen Banana	: 1 kg
ix)	Mustard (de-oiled cake)	: 2 kg
x)	Banyan tree & Banana soil	: 2 kg each

Application Procedure/Implementation:

- Put all the ingredients in drum
- Stir it thirty times clock-wise and thirty times anti-clock wise (twice daily)
- Keep the drum closed.
- Enriched *Panchagavya* would become ready on 7th day.
- Mix 100 liters of water and stir it properly.
- Mix the mixture with vermin-compost or ash or fertile soil and spread it in the field before sowing.

During irrigation, make a hole in the drum (4" above the base) and set it so that dripping from drum starts and whole field will be nourished. By its application at the time of sowing and during irrigations, soil can maintain remain fertile.

EM – Technology in organic farming

EM or effective microorganisms are a consortium culture of different effective microbes, commonly occurring in nature. Most important among them are: N_2-fixers, P-solubilisers, photosynthetic microorganisms lactic acid bacteria, yeasts, plant growth promoting rhizobacteria and various fungi and actinomycetes. In this consortium, each microorganism has its own beneficial role in nutrient recycling plant protection and soil health and fertility enrichment.

How to use EM

- Procurement of primary EM available in market.
- Preparation of secondary EM - to be carried by the farmer by using the primary EM.

EM-1 formulation

This is used for seed treatment, soil enrichment and spray in field after emergence of seedlings.

- Dissolve 5 kg jaggery (chemical free) in 100 lit of water
- Add 5 lit of EM
- Mix thoroughly and pour into a plastic carboy. Seal the carboy and allow to ferment for 7 days.
- Dilute this solution in a ratio of 1:100 and spray over soil or crop residue. For seed treatment soak the seeds in this dilute solution.

EM-5 formulation for control of insects and pests

- Dissolve 100 g jaggery (chemical free) in 600 ml of water
- Add 100 ml each of natural vinegar, wine or brandy and EM
- Mix thoroughly and transfer the contents in a plastic bottle or carboy. Seal the container and allow fermenting for 5-10 days under shade.
- To increase the potency few cloves of garlic and chilly paste can also be added to this suspension before sealing the container.

- Release the gas daily
- Within 10 days the EM solution will be ready for use. This can be stored upto 3 months at normal room temperature in a cool and dry palce
- Dilute the contents in a ratio of 1:1000 and apply as foliar spray

Fermented Plant Extracts (FPE)

- Ground 2.3 kg of fresh green weeds to a coarse paste
- Dilute with 14 lit of water
- Add 420 ml of EM
- Transfer the contents to a plastic drum and with the help of a thick plastic sheet cover the drum and tie with a rope
- The drum should be filled up to the top, leaving very little space for air.
- Fermentation and gas formation process will start slowly
- Mix the contents at repeated intervals
- Finished FPE having s pH of 3.5 with pleasing smell will be ready in 5-10 days time
- Filter the solution through a cloth and collect the filtrate
- For spraying on soil dilute the FPE in a ratio of 1:1000 with fresh water
- For spraying on crops dilute the FPE in a ratio of 1:500 with fresh water
- Sparing should be done after germination of seeds in early morning hours once or twice a week

Application of EM formulations

At the time of land preparation – Dilute 5-10 litres of simple EM solution in 50-100 lit of water and sprinkle/spray over 0.1 ha of land, when soil is wet a day before sowing.

For seed treatment – Soak seeds for 5-6 hrs in 1 : 100 fold diluted EM solution and sow immediately.

As foliar/ soil spray – After seedling emergence, 1 : 1000 diluted EM solution or FPE should be sprayed at the rate of 500 lit per ha, 4-5 times at an interval of 7-10 days. In fast growing crops such as vegetables, spraying should be done twice a week. In transplanted crops 1 : 500 diluted FPE can be sprayed after 5 days of transplanting @ 750-1000 lit per ha.

For soil enrichment – For every 0.1 ha mix 100-150 kg Bokashi with crop residue and mix with soil just before sowing. Simple EM solution @ 5-10 lit can also be used as spray over this residue-Bokashi mix. Spraying the soil with 5-10 lit of FPE mixed in 500-1000 lit of water per ha also add to the fertility of the soil

EM- Bokashi

It is a type of compost prepared by fermentation of waste organic matter with the help of EM which is mainly used for improving the fertility status of the soil and fro enhancing the degradation of crop residue.

- Collect sufficient quantity of organic matter (such as rice bran, fish meal, animal waste etc.,) equivalent to 150 lit drum volume
- Mix 150 g of jaggery and 50 ml of EM in 15 lit of water
- Mix this solution with organic waste thoroughly in such a way that entire contents get uniformly moistened
- Transfer the contents in plastic bag and seal the bag
- To ensure the anaerobic conditions put this bag into another polythene bag and seal
- Allow the contents to ferment for 3-4 days in a cool shade place
- Bokashi will ready afire 4 days and stored up to 6 months in a air tight plastic bags
- For 0.1 ha mix 100-150 kg bokashi with sufficient quantity of finely chopped crop residue and mix with soil just before sowing. Spraying of 5-10 lit of 1:500 dilute simple EM solutions over this mixture can further boost the degradation process.

Biogas slurry applications

The following are the different methods of applying bio digested slurry as manure:

a) Air dried biogas slurry can be applied by spreading on the agricultural land at least one week before sowing the seeds or transplanting the seedlings.

b) The liquid slurry can be mixed directly with the running water in irrigation canal which will enable spreading of the slurry uniformly in the cropped area or in cultivation land

c) Biogas slurry can also be coated on the seeds prior to sowing. This acts as insecticide and prevents seeds or plants from insect attack. This helps in early germination and healthy growth of seedlings.

d) The digested slurry is fed through the channel, flowing over a layer of green or dry leaves and filtered in the bed. The water from the slurry filters down which can be reused for preparing another fresh dung slurry. The semi-solid slurry can be transported easily as it was in the consistency of fresh dung and used for top dressing of crops like sugarcane and potato.

e) Biodigested slurry is also being used to fish culture, which acts as a supplementary feed. On an average, 15-25 litres of wet slurry can be applied per day in a 1200 sq m pond. Slurry mixed with oil cake or rice-bran in the 2:1 ratio increases the fish production remarkably. In general, organic manures about 10t/ha in the form of FYM or compost or biodigested slurry is recommended to be applied once in three years to maintain the organic content of soil, besides providing nitrogen, phosphorus and potassium in the form of organic fertilizers to the crop.

f) The digested slurry, if mixed with Azospirillum, KMB and PSM @ 200ml each/acre ensures increase of yield minimum 30% over slurry alone.

Conservation Tillage

The objective of conservation tillage is to provide a means of profitable crop production while minimizing soil erosion due to wind and/or water. The emphasis is on soil conservation, but conserving soil moisture, energy, labour, and even equipment provides additional benefits. To be considered conservation tillage, the system must provide conditions that resist erosion by wind, rain, and flowingwater. Such resistance is achieved either by protecting the soil surface with crop residues or growing plants or by maintaining sufficient surface roughness or soil permeability to increase water filtration and thus reduce soil erosion.

Conservation tillage is often defined as any crop production system that provides either a residue cover of at least 30% after planting to reduce soil erosion due to water or at least 1,000 pounds per acre of flat, small-grain residues (or the equivalent) on the soil surface during the critical erosion period to reduce soil erosion due to wind.

The term *conservation tillage* represents a broad spectrum of tillage systems. However, maintaining an effective amount of plant residue on the soil surface is the crucial issue, which is why the Natural Resources Conservation Service (NRCS) has replaced conservation tillage with the term *crop residue management*. This term refers to a philosophy of year-round management of residue to maintain the level of cover needed for adequate control of erosion.

Adequate erosion control often requires more than 30% residue cover after planting. Other conservation practices or structures may also be required. Some of the conservation tillage systems are described here.

No-Till

With no-till, the soil is left undisturbed from harvest to seeding and from seeding to harvest. The only "tillage"is the soil disturbance in a narrow band created by a row cleaner, coulter, seed furrow opener, or other device attached to the planter or drill. Many no-till planters arenow equipped with row cleaners to clear row areas of residue.

No-till planters and drills must be able to cut residue and penetrate undisturbed soil. In practice, a tillage system that leaves more than 70% of the surface covered by crop residue is considered to be a no-till system.

Strip-Till

Strictly speaking, a no-till system allows no operations that disturb the soil other than planting or drilling. On some soils, including poorly drained ones, the no-till system is sometimes modified by the use of a strip tillageoperation, typically in the fall, to aid soil drying and warming in the spring. This system is called strip-till. Itis considered a category of no-till, as long as it leaves the necessary amount of surface residue after planting.

Strip-till is sometimes done along with the fall applicationof anhydrous ammonia, dry fertilizer, or both. This usually involves using a mole knife, which is designed to shatter and lift soil as it places fertilizer. A closing apparatus, usually disk blades run parallel to the row, pulls soil into the row. In some cases a rolling cage is used to firm the strip and break up clods. This process creates a small, elevated strip called a berm.

One benefit of strip-till, compared to no-till, is accelerated soil warming that results from removing residue and disturbing the soil in the berm. Planting takes place asclose as possible to the center of the berm, which has usually "melted down" by spring to be little higher than the soil between the rows. The width of the strip-till implementis usually matched to the planter width, and the use of RTK-directed autosteer greatly assists the strip-till and planting processes. Maintenance of interrow residue helpsto provide the benefits of a no-till system, while the uncovered soil near the seed row reduces the negative effects of cold, wet soils often found in no-till. The advantages of strip-till over no-till are thus most likely to be seen incold, wet springs.

Agro-industrial wastes

Agricultural residues are rich in bioactive compounds. These residues can be used as an alternate source for the production of different products like biogas, biofuel, mushroom, and tempeh as the raw material in various researches and industries. The use of agro-industrial wastes as raw materials can help to reduce the production cost and also reduce the pollution load from the environment. Agro-industrial wastes are used for manufacturing of biofuels, enzymes, vitamins, antioxidants, animal feed, antibiotics, and other chemicals through solid state fermentation (SSF).

Types of agro-industrial wastes

Agricultural residues

Crop residue, traditionally considered as "trash" or agricultural waste, is increasingly being viewed as a valuable resource. Corn stalks, corn cobs, wheat straw, paddy straw and other leftovers from grain production are now being viewed as a resource with economic value. If the current trend continues, crop residue will be a "co-product" of grain production where both the grain and the residue have significant value. The emergence ofcrop residue as a valuable resource has evolved to the point where there are competing uses forit. "Crop residue, in general, are parts of plants left in the field after crops have been harvested and threshed or left after pastures are grazed. These materials have at times been regarded as waste materials that require disposal but it has become increasingly realized that they are important natural resources and not wastes

When Crop Residue Incorporated in to Field it Helps in Improving

- Organic carbon and N content in soil.
- Acts as a buffer in soil against rapid change in soil pH.
- Reclamation and management of saline and alkaline soil.
- Incorporation of organic materials the pH and ESP in the soil and improved crop yield.
- Acts as a reservoir for plant nutrients and prevents leaching of elements, which are essential for plant growth.
- The incorporation of straw along with application of FYM, reduce the bulk density of soil and increases the porosity of the soil.
- Provide energy for growth and activities of microbes.
- Improving soil and water conservation and sustaining soil fertility and enhancing crop yields

- Raised the soil temperature in winter and lowered it in summer season.

Crop residue management

- Mulching
- Animal feed
- Compost preparation

Mulching

Mulching means covering the soil surface with any material such as organic wastes, plastic, polythene etc. the organic wastes used for mulching including crop stubbles, straw, coir pith, groundnut shell, husk etc. These wastes at 5-10 t/ha are spread on the soil surface to a thickness of 5-10 cm.

Animal feed

In many parts of the world (particularly in areas with low yield potential), crop residues are essential for feeding to animals. Animals are left to graze freely on harvested fields or straw is collected off the land and taken to the pen to feed livestock and prepare bedding. This scenario, if not managed sustainably, often results in deteriorating soil fertility; nutrients are removed from the land and not returned and the bare soil is exposed to wind and water erosion. Some nutrients may be returned to the soil through the manure left by grazing animals; however, most of the nitrogen is lost to the air.

Composting

Composting is allowing organic material to decompose under more or less controlled conditions to produce a stabilized product that can be used as a soil amendment

Crop Rotation

Crops are changed year by year in a planned sequence. Crop rotation is a common practice on sloping soils because of its potential for soil saving. Rotation also reduces fertilizer needs, because alfalfa and other legumes replace some of the nitrogen corn and other grain crops remove.Crop rotation—especially an extended crop rotation using three or more crops—is an age-old sustainable farming practice capable of maintaining crop yields.

- Non –leguminous crops should be followed by leguminous crops and vice-versa, eg. green gram – wheat / maize. If preceding crops are legume or non-legume grown as intercrops or mixed crops, the succeeding crop may be legume or non legume or both.
- Restorative crops should be followed by exhaustive or non-restorative crops.eg. sesame – cowpea / green gram / blackgram / groundnut.
- Leaf shedding crop should be followed by non-leaf shedding or less exhaustive crops.eg. pulses / cotton – wheat / rice.
- Green manuring crop should be followed by grain crops.eg. dhaincha - rice, green gram/ cowpea – wheat / maize.
- Highly fertilized crops should be followed by non-fertilised crop.eg. maize - black gram/gourds.
- Perennial or long duration crops should be followed by seasonal /restorative crops. eg. napier / sugarcane - groundnut /cowpea /green gram.
- Fodder crops should be followed by field or vegetable crops. eg. maize + cowpea-wheat/potato/cabbage/onion.
- Multicut crops should be succeeded by the seed crops. eg. green gram/maize.
- Ratoon crops should be followed by deep rooted restorative crops. eg. sugarcane/jowar-pigeonpea/Lucerne/cowpea.
- Fouling crops should be followed by cleaning crops.eg. jowar /maize potato/ groundnut.
- Cleaning crops should be followed by nursery crops. eg. potato/ colocasia/ turmeric / beet/ carrot-rice nursery/ onion nursery/ tobacco nursery/ vegetable nursery.
- Deep rooted crops should be succeeded by shallow rooted crops. eg. cotton/ castor/ pigeonpea – potato / lentil /green gram etc.

- Deep tillage crops should be followed by zero or minimal tillage crops. eg. potato / radish / sweet potato/sugarcane - black gram/ green gram/green manuring crops.
- Dicot crops should be followed by monocot crops. eg, potato / mustard / groundnut / pulses – rice / wheat / sugarcane / jowar or dicot + Monocot crops should be followed by dicot + monocot or either dicot or monocot crops.
- Stiff stubble leaving crops should be followed by minimum intercultivation requiring crops. eg. sugarcane / sorghum/cotton /pigeonpea- fodder crops.
- The crops of wet (anaerobic) soil should be followed by the crops of dry (aerobic). eg. rice-Bengal gram/Lathyrus/pulses/ oilseeds. The tendency to build up difficult-to-control weeds becomes less in such rotation than in continuous wet land rice culture.
- The crops that are susceptible to soil-borne pests and pathogens should be followed by tolerant / break / trap crops. eg. sugarcane-marigold for pathogenic nematodes, tomato / brinjal / tobacco / potato-rice / pulses for Orobanche, jowar-castor for Striga and berseem-oats for Cuscuta.
- The crops with problematic weeds (weeds that are difficult to distinguish at any one stage of crop, may be seedling or seed stage) should be followed by cleaning crops/ multicut crops / other dissimilar crops or varieties. eg. wheat-wet rice for *Phalaris minor*, berseem-potato/boro rice for *Cichorium intybus*, mustard early potato for *Cleome viscosa*, rice-jute / sugarcane/vegetable/maize + cowpea for *Echinochloa crusgalli*, jute- multicut fodder/vegetable or *Corchorus acutangulus*.
- Pasture crops should be followed by fodder or seed crop. eg. para grass – maize + cowpea / cowpea / rice bean / tetrakalai for seed.
- Silage / hay / cleaning crops should be followed by seed crops. eg. maize / groundnut - onion, cowpea / jowar for seed crops.

- Crops with the same symbiotic / associated microbes should be followed by common host crops, such as,

Rhizobium meliloti - lucerna sweet clover, fenugreek

R. trifolii - berseem, Persian clover

R. leguminosoarum - peas, lentil, lathyrus

R. phaseoli - beans, green gram, pillipesara, black gram

R. lupine - Lupines

R. japonicum - cowpea, pigeonpea, guar, sunhemp, bengal gram, soybean,

The rotational use of crop varieties, and cultural practices in addition to rotational cropping provides more and assured benefits than that of adopting only crops or land rotation.

Chapter 7

Organic Crop Protection

From an ecological perspective all organisms are part of nature, irrespective of what they do. To a farmer, all organisms that reduce the yields of their crops are considered pests, diseases or weeds. Insects, birds or other animals are also pests whenever they cause damage to crops or stored produce. Fungi, bacteria and viruses are also recognized as disease causing organisms when they lead to conditions that interrupt or modify the vital functions of growing plants or stored produce. All unwanted plants that grow within crops and compete with them for nutrients, water and sunlight are considered weeds. Such plants can also be hosts for pests and diseases.Presence of these organisms in crop fields is not a problem until their numbers increase beyond a level where they attack, and cause substantial reductionin field crop yields or quality of harvested and stored produce.

The organic approach to plant pest, plant disease and weed management makes reference to the four principles of organic agriculture: the principle of health, the principle of ecology, the principle of fairness, and the principle of care. Generally, organic farmers aim at sustaining and enhancing the health of their soils, plants, animals, humankind and—in the widest sense—the planet. The health of individuals and communities cannot be separated fromthe health of ecosystems. Therefore, by providing healthy soils and a diversified natural environment, farmers are able to produce healthy crops that foster the health of animals and people.

Healthy plants are also able to resist and tolerate physiological disruption and damage from disease-causing organisms and pests.

Thus, organic farmers aim at optimizing the growing conditions for their crops to make them strong and competitive. At the same time they encourage natural control mechanisms to prevent pests, diseases and weeds to develop in a way that they cannot damage the crops. They, therefore, give priority to preventive measures to prevent and limit the spread of infections, instead of relying on direct control measures. Direct control measures are mainly applied when pathogens have already developed.

Insect management

Insect management presents a challenge to organic farmers. Insects are highly mobile and well adapted to farm production systems and pest control tactics. Organic farmers aim at creating diverse and ecologically stable farming systems and enhancing the plants' natural defence mechanisms. The goal is a robust,healthy crop plant. They give first priority to prevention of pest and disease development instead of direct control measures. At the same time, this approach enhances biodiversity, protects natural resources and favours natural control mechanisms.

Components of organic insect management

The following components may be included in organic method of pest management

1. Preventive practices
2. Habitat diversification
3. Mechanical methods of pest management
4. Physical methods of pest management
5. Use of insect pheromones
6. Biological control of pests
7. Use of synthetic organics permissible for use in organic agriculture
8. Using farmers wisdom in organic farming
9. Use of plant products / botanicals

1. Preventive practices

1) *Selection of adapted and resistant varieties*

- Choose varieties which are well adapted to the local environmental conditions (temperature, nutrient supply, pests and disease pressure), as it allows them to grow healthy and makes them stronger against infections of pests and diseases.

2) *Selection of clean seed and planting material*

- Use planting material from safe sources.
- Planting physically sound seed is also important. In crops such as flax, rye and pulses, a crack in the seed coat may serve as an entry point for soil-borne microorganisms that rot the seed once it is planted.

3) *Use of suitable cropping systems*

- Mixed cropping systems: can limit pest pressure as the pest has less host plants to feed on and more beneficial insect life in a diverse system.
- Crop rotation: reduces the chances of host for pests and increases soil fertility.

4) *Organic matter addition*

- Stabilises soil structure and thus improves aeration and infiltration of water.
- Supplies substances which strengthen the plant's own protection mechanisms.

5) *Application of suitable soil cultivation methods*

- Facilitates the decomposition of infected plant parts.
- Regulates weeds which serve as hosts for pests and diseases.

6) *Conservation and promotion of natural enemies*

- Provide an ideal habitat for natural enemies to grow and reproduce.
- Avoid using products which harm natural enemies.

7) *Selection of optimum planting time and spacing*

- Most pests attack the plant only in a certain life stage; therefore it's crucial that this vulnerable life stage doesn't correspond with the period of high pest density and thus that the optimal planting time is chosen.

8) *Crop rotation*

Crop rotation is central to all sustainable farming systems. It is an extremely effective way to minimize most pest problems while maintaining and enhancing soil structure and fertility. Diversity is the key to a successful crop rotation program. It involves:

- Rotating early-seeded, late-seeded and fall-seeded crops
- Rotating between various crop types, such as annual, winter annual, perennial, grass and broadleaf crops; each of these plant groups has specific rooting habits, competitive abilities, nutrient and moisture requirements. (True diversity does not include different species within the same family - for example, wheat, oats and barley are all species of annual cereals.)
- Incorporating green manure crops, into the soil to suppress pests, disrupt their life cycles and to provide the additional benefits of fixing nitrogen and improving soil properties
- Managing the frequency with which a crop is grown within a rotation
- Maintaining the rotation's diversified habitat, which provides parasites and predators of pests with alternative sources of food, shelter and breeding sites
- Planting similar crop species as far apart as possible. Insects such as wheat midge and Colorado potato beetle, for example, are drawn to particular host crops and may over winter in or near the previous host crops. With large distances to move to get to the successive crop, the insects' arrival may be delayed. The number that find the crop may be reduced as well.

9) *Forecasting*

Producers should pay attention to the forecasts for various pest and disease infestations for each crop year. Maps of these forecasts are usually available for many of the major destructive insects such as grasshoppers and wheat midge, as well as some diseases. Agro meteorological warning and forecast can help in this way.

2. Habitat diversification

Organic farmers make use of habitat management practices like conservation biocontrol, intercropping and trap cropping to encourage natural enemies of pest and disease causing organisms. For improving habitats, only non-host plant species of important pests and diseases should be chosen (e.g. avoiding plant species of the Brassica family in case of Turnip mosaic virus or Solanaceae family in case of late blight).

Habitat management also includes the adjustment of the environment around and within the field to improve air circulation (e.g. pruning trees and bushes limiting the air circulation in the main wind direction and avoiding densecrops). In contrast to pest management, the possibilities to reduce diseases by habitat management are more limited. However, the practices are also helpful to stabilize the entire system.

Intercropping system

Intercropping system has been found favourable in reducing the population and damage caused by many insect pests due to one or more of the following reasons.

- Pest outbreak less in mixed stands due to crop diversity than in sole stands
- Availability of alternate prey
- Decreased colonization and reproduction in pests
- Chemical repellency, masking, feeding inhibition by odours from non-host plants.
- Act as physical barrier to plants.

Table 11. Effect of intercropping system on pest levels

Sl.No.	Crop		Pest reduced
	Sole crop	Intercrop	
1.	Sorghum	Red gram	Earhead bug
2.	Sorghum	Cowpea	*Chilo partellus stem borer*
3.	Pigeon pea	Sorghum	Leaf hopper Empoasca kerri

5.	Green gram	Sorghum	*E. kerri* Leaf hopper
5.	Ground nut	Sorghum	*E.kerri* Leaf hopper
6.	Pigeon pea	Sorghum	*H. armigera*
7.	Chickpea	Wheat, Mustard or Safflower	*H.armigera*
8.	Sugarcane	Greengram, Blackgram	Early shoot borer

Trap cropping

Crops that are grown to attract insects or other organisms like nematodes to protect target crops from pest attack. This is achieved by

- Either preventing the pests from reaching the crop or
- Concentrating them in a certain part of the field where they can be economically destroyed

Fig.18. Castor as a trap crop in the field

Ecological engineering

Ecological engineering for pest management has recently emerged as a paradigm for considering pest management approaches that rely on the use of cultural techniques to effect habitat manipulation and to enhance biological control. The cultural practices are informed by ecological knowledge rather than on high technology approaches such as synthetic pesticides and genetically engineered crops.

Natural enemies may require

- Food in the form of pollen and nectar for adult natural enemies
- Shelters such as overwintering sites, moderate microclimate etc.
- Alternate hosts when primary hosts are not present.

The application of ecological engineering for pest management includes use of cultural practices, usually based on vegetation management, to enhance biological control. Pest problems have increased tremendously due to monoculture, overlapping of crops, dense cropping, and availability of preferred host. Alternative to chemical method is biological control and in ecological engineering concept, growing flowering plants is key component to provide resources such as nectar and pollen to natural enemies to promote biological control. It includes attractant plants to attract the natural enemies, repellent plants to repel the pests, trap plants to attract and trap the crop pests, barrier/guard plants to prevent the entry of pests. It also includes trap crops that divert pests away from crops and changing monocultures to polycultures in agroecosystem.

Ecological engineering for pest management mainly focuses on increasing the abundance, diversity and function of natural enemies in agricultural habitats by providing refugees and alternate or supplementary food resources. Ecological engineering for pest management has roots in traditional forms of mixed farming systems

Ecological engineering for pest management – Above ground

- Raise the flowering plants / compatible cash crops along the orchard border by arranging shorter plants towards main crop and taller plants towards the border to attract natural enemies as well as to avoid immigrating pest population
- Grow flowering plants on the internal bunds inside the orchard
- Not to uproot weed plants those are growing naturally like *Tridax procumbens, Ageratum* sp, *Alternanthera* sp etc. which act as nectar source for natural enemies,
- Not to apply broad spectrum chemical pesticides, when the Pest diseases ratio is favourable. The plant compensation ability should also be considered before applying chemical pesticides.

Ecological engineering for pest management – Below ground

- Keep soils covered year-round with living vegetation and/or crop residue.
- Add organic matter in the form of farm yard manure (FYM), Vermicompost, crop residue which enhance below ground biodiversity.
- Reduce tillage intensity so that hibernating natural enemies can be saved.
- Apply balanced dose of nutrients using biofertilizers.
- Apply mycorrhiza and plant growth promoting rhizobacteria (PGPR)
- Apply *Trichoderma* spp. and *Pseudomonas fluorescens* as seed/ seedling/planting material, nursery treatment and soil application (if commercial products are used, check for label claim. However, biopesticides produced by farmers for own consumption in their fields, registration is not required).

Due to enhancement of biodiversity by the flowering plants, parasitoids and predators (natural enemies) number also will increase due to availability of nectar, pollen, fruits, insects, etc. The major predators are a wide variety of spiders, ladybird beetles, long horned grasshoppers, *Chrysoperla*, earwigs, etc.

3. Cultural practices

Crop rotation

Sustainable systems of agricultural production are seen in areas where proper mixtures of crops and varieties are adopted in a given agro-ecosystem. Monocultures and overlapping crop seasons are more prone to severe outbreak of pests and diseases. For example growing rice after groundnut in garden land in puddled condition eliminates white grub.

Organic manure

Application of press mud in groundnut @ 12.5 t/ha had a better influence on leaf miner with lower leaflet damage at 38.84 per cent and 2.48 larval numbers per plant during summer 1991. It was 34.93 per cent and 2.72 numbers during kharif, 1991. Farm yard manure, *Azospirillum* and Phosphobacteria has no significant influence on the control of leaf hopper and fruit borer in bhendi. The incidence of paddy plant and leafhopper was low in *Azospirillum* combined with farmyard manure. Application of organic manure lowered the rice gall midge incidence (5.28%)

Depth and Timing of Seeding

Optimum seeding depth is also important. Deep seeding in cold soils may result in seedling blights and damping-off, especially in pulses and small-seeded crops. Seeding depth should generally be no deeper than required for quick germination and even emergence. Variables include seed size, soil type and moisture conditions. If the soil is loose before seeding, a packing operation will firm up the soil and bring moisture closer to the surface.

For most crops, seeding should ideally be done when the soil is warm enough for rapid germination. Seeds that remain ungerminated in cool soil are more susceptible to damage by insects such as wireworms.

Water Management

Irrigation has both direct and indirect effects on pest insects. Insect populations can decrease if overhead sprinklers knock insects off plants or raise microenvironment humidity enough to encourage insect disease caused by bacteria or fungi. Because irrigation methods vary considerably (whether drip, overhead sprinkler, or flood irrigation), theimpact of irrigation on insects also varies. Pestinsect populations can increase if irrigated plants are lusher and more attractive than surrounding plants. Likewise, plants stressed by drought can be more attractive to insectpests or less tolerant to their feeding. The need for irrigation is dictated by crop growth and weather rather than the need for insect control. But when there is some flexibility in irrigation scheduling, a farmer should think about irrigation as a tool for suppressing pestinsects. Several naturally occurring insect pathogens, especially insect-pathogenic fungi, provide effective pest suppression when high humidity microenvironments are created.

Tillage

Tillage practices affect both subterranean and foliar insect pests. Infrequent disturbance of soils in natural systems preserves food webs and diversity of organisms and habitats. Theregular disturbance of agricultural soils disrupts ecological linkages and allows adapted pest species to increase without the dampening effects of natural controls. Nevertheless, tillage can also destroy insects overwintering in the soil as eggs, pupae, or adults, and reduce pest problems. Organic producers usually rely on tillage to control weeds and to prepare the soil for planting. Research is being conducted on methods and equipment that may allow for the reduction of tillage in organic systems.

Some practices to reduce tillage in organic systems include zone tillage, ridge tillage, and including a perennial or sod-producing cropin the rotation. Reduction of tillage alters pest insect dynamics considerably. Thrips cause fewer problems in reduced-till systems. Ground-dwelling predators, such as ground beetles that prey on pest insects, can increase. However, cutworm and slug problems can also increase where tillage is reduced. The degree of pest population shifts between atilled and reduced-tillage system cannot be reliably predicted. Species shifts will occur and should be carefully monitored. Tillage is not likely to have any significant effect on most common foliar-feeding insect pests.

Summer ploughing

Summer ploughing is an important cultural practice for pest control. When the land is ploughed, the inactive stages of pests like egg masses, larvae and pupae present within 5–10 cm surface of the soil get exposed. They are killed due to the intense heat of summer and are also eaten away by predatory birds.

Companion Planting

The companion planting approach is based on the theory that various plants grown in close proximity to the crop plant will repel or kill pest insects. Studies to date have not shown this approach to be effective. Note that companion planting is not the same as intercropping, which may be a valuable tool in attracting beneficial insects.

Biofumigation

Biofumigation is based on incorporating fresh plant mass into the soil, which will release several substances (mainly isothiocyanates) able to suppress soilborne pests. Good effects of biofumigation are also seen against *Sclerotinia* and *Phytium* disease, damping off (*Rhizoctonia solani*) and bacteria rot (*Erwinia* sp.). Plants from the Brassica family (mustard, radish, etc.) release large amounts of these toxic substances in the soil while decomposing and are considered the best material for biofumigation. Different mustards (e.g. *Brassica juncea var. integrifoliar Brassica juncea var. juncea*) should be used as intercrops on infested fields. As soon as mustards are flowering, they are cut and instantly incorporated into the soil by hoeing or ploughing. While incorporated plant parts are decomposing in a moist soil, nematicidal compounds are produced that kill nematodes. During the decomposing process, phytotoxic substances are released too. They can kill weed seeds, but they can also affect crop plants. Therefore, a new crops hould only be planted or sown two weeks after incorporating plant material into the soil.

4. Physical control

Modification of physical factors in the environment to minimise (or) prevent pest problems. Use of physical forces like temperature, moisture, etc. in managing the insect pests.

A. Manipulation of temperature

1. Sun drying the seeds to kill the eggs of stored product pests.
2. Hot water treatment (50-55°C for 15 min) against rice white tip nematode.
3. Flame throwers against locusts.
4. Burning torch against hairy caterpillars.
5. Cold storage of fruits and vegetables to kill fruit flies (1-2°C for 12-20 days).

B. Manipulation of moisture

1. Alternate drying and wetting rice fields against BPH.
2. Drying seeds (below 10% moisture level) affects insect development.
3. Flooding the field for the control of cutworms.

C. Manipulation of light

1. Treating the grains for storage using IR light to kill all stages of insects (eg.) Infrared seed treatment unit.
2. Providing light in storage go downs as the lighting reduces the fertility of Indian meal moth, *Plodia.*
3. Light trapping.

D. Manipulation of air

1. Increasing the CO_2 concentration in controlled atmosphere of stored grains to cause asphyxiation in stored product pests.

E. Use of irradiation

Gamma irradiation from CO^{60} is used to sterilize the insects in laboratory which compete with the fertile males for mating when released in natural condition. (eg.) cattle screw worm fly, *Cochliomyia hominivorax* control in Curacao Island by E.F.Knipling

Use of greasing material

Treating the stored grains particularly pulses with vegetable oils to prevent the oviposition and the egg hatching. eg., bruchid adults.

Bird perches

'T' shaped bird perches should be erected in the field at the rate of 15–20 per hectare. They should be placed one foot above the crop canopy. These perches serve as resting places for the birds which feast on the larvae they find in the field. Mix rice with the blood of a chicken, make it into pellets and broadcast these in the field. The smell of blood and rice attracts predatory birds to the perches in the field from where they pick up the swarming caterpillars.

Rope method

The field should be filled with water up to a height of 5 cm. One litre of kerosene should be mixed with 25 kg of sand and strewn in the field. Later, a string should be dragged over the surface of the leaves vigorously so that the caterpillars fall into the water. The caterpillars are killed by the kerosene present in the water. Later, the water should be drained to remove the dead caterpillars. The field should be dried and then freshly irrigated. This method should be used only during the vegetative stages of the crop.

Use of effigies

A human-like figure, made of paddy straw and wearing a white dress (@ two effigies per hectare) kept in the field at milky to grain filling stage, will scare away the birds.

Use of visible radiation: Yellow colour preferred by aphids, cotton whitefly: yellow sticky traps.

Use of Abrasive dusts

1. Red earth treatment to red gram: Injury to the insect wax layer.
2. Activated clay: Injury to the wax layer resulting in loss of moisture leading to death. It is used against stored product pests.
3. Drie-Die: This is a porous finely divided silica gel used against storage insects.

5. Mechanical control

Use of mechanical devices or manual forces for destruction or exclusion of pests.

Mechanical destruction: Life stages are killed by manual (or) mechanical force.

Manual Force

- Hand picking the caterpillars
- Beating: Swatting housefly and mosquito
- *Sieving and winnowing:* Red flour beetle (sieving) rice weevil (winnowing)

- **Shaking the plants:** Passing rope across rice field to dislodge caseworm and shaking neem tree to dislodge June beetles
- **Hooking :** Iron hook is used against adult rhinoceros beetle
- **Crushing :** Bed bugs and lice
- **Combing :** Delousing method for Head louse
- **Brushing :** Woolen fabrics for clothes moth, carpet beetle.

Mechanical force

- **Entoletter :** Centrifugal force - breaks infested kernels - kill insect stages - whole grains unaffected - storage pests.
- **Hopper dozer :** Kill nymphs of locusts by hoarding into trenches and filled with soil.
- **Tillage implements :** Soil borne insects, red hairy caterpillar.
- **Mechanical traps :** Rat traps of various shapes like box trap, back break trap, wonder trap, Tanjore bow trap.

Mechanical exclusion

Mechanical barriers prevent access of pests to hosts.

- **Wrapping the fruits :** Covering with polythene bag against pomegranate fruit borer.
- **Banding :** Banding with grease or polythene sheets - Mango mealybug.
- **Netting :** Mosquitoes, vector control in greenhouse.
- **Trenching :** Trapping marching larvae of red hairy caterpiller.
- **Sand barrier :** Protecting stored grains with a layer of sand on the top.
- **Water barrier :** Ant pans for ant control.
- **Tin barrier :** Coconut trees protected with tin band to prevent rat damage.
- **Electric fencing :** Low voltage electric fences against rats.

Appliances in controlling the pests

1. **Light traps :** Most adult insects are attracted towards light in night. This principle is used to attract the insect and trapped in a mechanical device.

a) Incandescent light trap : They produce radiation by heating a tungsten filament. The spectrum of lamp include a small amount of ultraviolet, considerable visible especially rich in yellow and red. (eg.) Simple incandescent light trap, portable incandescent electric. Place a pan of kerosenated water below the light source.

b) Mercury vapour lamp light trap : They produce primarily ultraviolet, blue and green radiation with little red. (eg.) Robinson trap. This trap is the basic model designed by Robinson in 1952. This is currently used towards a wide range of Noctuids and other nocturnal flying insects. A mercury lamp (125 W) is fixed at the top of a funnel shaped (or) trapezoid galvanized iron cone terminating in a collection jar containing dichlorvos soaked in cotton as insecticide to kill the insect.

c) Black light trap : Black light is popular name for ultraviolet radiant energy with the range of wavelengths from 320-380 nm. Some commercial type like Pest-O-Flash, Keet-O-Flash are available in market. Flying insects are usually attracted and when they come in contact with electric grids, they become elactrocuted and killed.

2. **Pheromone trap :** Synthetic sex pheromones are placed in traps to attract males. The rubberised septa, containing the pheromone lure are kept in traps designed specially for this purpose and used in insect monitoring/mass trapping programmes. Sticky trap, water pan trap and funnel type models are available for use in pheromone based insect control programmes.
3. **Yellow sticky trap :** Cotton whitefly, aphids, thrips prefer yellow colour. Yellow colour is painted on tin boxes and sticky material like castor oil/vaseline is smeared on the surface. These insects are attracted to yellow colour and trapped on the sticky material.
4. **Bait trap :** Attractants placed in traps are used to attract the insect and kill them. (eg.) Fishmeal trap : This trap is used against sorghum shoot fly. Moistened fish meal is kept in

polythene bag or plastic container inside the tin along with cotton soaked with insecticide (DDVP) to kill the attracted flies.

5. **Pitfall trap** helps to trap insects moving about on the soil surface, such as ground beetles, collembola, spiders. These can be made by sinking glass jars

   (or) metal cans into the soil. It consists of a plastic funnel, opening into a plastic beaker containing kerosene supported inside a plastic jar.

6. **Probe trap :** Probe trap is used by keeping them under grain surface to trap stored product insect.

7. **Emergence trap :** The adults of many insects which pupate in the soil can be trapped by using suitable covers over the ground. A wooden frame covered with wire mesh covering and shaped like a house roof is placed on soil surface. Emerging insects are collected in a plastic beaker fixed at the top of the frame.

8. **Indicator device for pulse beetle detection postharvest pest trap:** A new cup shaped indicator device has been recently designed to predict timely occurrence of pulse beetle *Callosobruchus* spp. This will help the farmers to know the correct time of emergence of pulse beetle. This will help them in timely sun drying which can bill all the eggs.

6. Biological control

The successful management of a pest by means of another living organism (parasitoids, predators and pathogens) that is encouraged and disseminated by man is called biological control. In such programme the natural enemies are introduced, encouraged, multiplied by artificial means and disseminated by man with his own efforts instead of leaving it to nature.

Techniques in biological control

Biological control practices involve three techniques *viz.*, Introduction, Augmentation and Conservation.

1. **Introduction or classical biological control:** It is the deliberate introduction and establishment of natural enemies to a new locality where they did not occur or originate naturally. When natural enemies are successfully established, it usually continues to control the pest population.
2. **Augmentation**: It is the rearing and releasing of natural enemies to supplement the numbers of naturally occurring natural enemies. There are two approaches to augmentation.
   a. **Inoculative releases**: Large number of individuals are released only once during the season and natural enemies are expected to reproduce and increase its population for that growing season. Hence control is expected from the progeny and subsequent generations and not from the release itself.
   b. **Inundative releases**: It involves mass multiplication and periodic release of natural enemies when pest populations approach damaging levels. Natural enemies are not expected to reproduce and increase in numbers. Control is achieved through the released individuals and additional releases are only made when pest populations approach damaging levels.
3. **Conservation: Conservation** is defined as the actions to preserve and release of natural enemies by environmental manipulations or alter production practices to protect natural enemies that are already present in an area or non use of those pest control measures that destroy natural enemies.

Important conservation measures are

- Use selective insecticide which is safe to natural enemies.
- Avoidance of cultural practices which are harmful to natural enemies and use unfavourable cultural practices
- Cultivation of varieties that favour colonization of natural enemies
- Providing alternate hosts for natural enemies.
- Preservation of inactive stages of natural enemies.
- Provide pollen and nectar for adult natural enemies

Parasite: A parasite is an organism which is usually much smaller than its host and a single individual usually doesn't kill the host. Parasite may complete their entire lifecycle (eg. Lice) or may involve several host species. Or Parasite is one, which attaches itself to the body of the other living organism either externally or internally and gets nourishment and shelter at least for a shorter period if not for the entire life cycle. Theorganism, which is attacked by the parasites, is called hosts.

Parasitism: Is the phenomena of obtaining nourishment at the expense of the host to which the parasite is attached.

Parasitoid: is an insect parasite of an arthropod, parasitic only in immature stages,destroys its host in the process of development and free living as an adult. Eg: Braconid wasps

Qualities of a Successful Parasitoid in Biological Control Programme

A parasitoid should have the following qualities for its successful performance.It

1. Should be adaptable to environmental conditions in the new locally
2. Should be able to survive in all habitats of the host
3. Should be specific to a particulars sp. of host or at least a narrowly limited range of hosts.
4. Should be able to multiply faster than the host
5. Should be having more fecundity
6. Life cycle must be shorter than that of the host
7. Should have high sex ratio
8. Should have good searching capacity for host
9. Should be amenable for mass multiplication in the labs
10. Should bring down host population within 3 years
11. Should be free from hyperparasitoids
12. There should be quick dispersal of the parasitoid in the locality

Trichogramma

Trichogramma is a bio-control agent against lepidopteron insect pests. It is available in the form of cards 20,000 live parasitoid eggs which have 90-96% hatching within 7-days of parasitization.

Chrysoperla

Chrysoperla is a bio-control agent against most of the insect pests. It is applied to crops @ 5000-10000 eggs/larvae per ha, 3-4 times at 15 days interval coinciding with egg laying young stages of pests. These are applied @ 3 cards per ha, 3-4 times at 15 days interval in various crops.

Predators and Predatism

A predator is one which catches and devours smaller or more helpless creatures by killing them in getting a single meal. It is a free living organism through out its life ,normally larger than prey and requires more than one prey to develop.

Insect predator qualities

- A predator generally feeds on many different species of prey , thus being a generalist or polyphagous nature
- A predator is relatively large compared to its prey, which it seizes and devours quickly
- Typically individual predator consumes large number of prey in its life time Eg: A single coccinellid predator larva may consume hundreds of aphids
- Predators kill and consume their prey quickly , usually via extra oral digestion
- Predators are very efficient in search of their prey and capacity for swift movements
- Predators develop separately from their prey and may live in the same habitat or adjacent habitats
- Structural adaptation with well developed sense organs to locate the prey

- Predator is carnivorous in both its immature and adult stages and feeds on the same kind of prey in both the stages
- May have cryptic colorations and deceptive markings

Eg. Praying mantis and Robber flies

Predatism

Based on the degree of usefulness to man, the predators are classified as on

1. Entirely predatory, Eg. lace wings, tiger beetles lady bird beetles except *Henosepilachna* genus
2. Mainly predator but occasionally harmful. Eg. Odonata and mantids occasionally attack honey bees
3. Mainly harmful but partly predatory. Eg. Cockroach feeds on termites. Adult blister beetles feed on flowers while the grubs predate on grass hopper eggs.
4. Mainly scavenging and partly predatory. Eg. Earwigs feed on dead decaying organic matter and also fly maggots. Both ways, it is helpful
5. Variable feeding habits of predator, eg: Tettigoniidae: omnivorous and carnivorous but damage crop by laying eggs.
6. Stinging predators. In this case, nests are constructed and stocked with prey, which have been stung and paralyzed by the mother insect on which the eggs are laid and then scaled up. Larvae emerging from the egg feed on paralyzed but not yet died prey. Eg. Spider wasps and wasps.

Table 12. Differences Between predator and a parasite

Predator	Parasite
Mostly a generalized feederexcepting lady bird beetles andhover flies which show somespecificity to pray	Exhibits host specialization and inmany cases the range of host speciesattacked is very much limited
Very active in habits	Usually sluggish one the host issecured.
Organs of low common senseorgans and mouth parts are welldevelop	Not very well developed and sometimes reduced even, Ovipositor well developed and oviposition specialized
Stronger, larger and usually moreintelligent than the prey	Smaller and not markedly moreintelligent than the host
Habitat is in dependent of that ofits prey	Habitat and environment is made anddetermined by that of the host
Life cycle long	Life cycle Short
Attack on the prey is casual andnot well planned	Planning is more evident
Seizes and devours the preyrapidly	Lives on or in the body of the hostkilling it slowly
Attack on prey is for obtaining foodfor the attacking predator itself,excepting in wasps which sting thecaterpillars to paralyze the andprovide them as food in the nestfor the young	It is for provision of food for the off spring
A single predatory may attackseveral hosts in a short period	A parasite usually completes development in a single host in mostcases

Beneficial insects

Management of honey bees for pollination

- Place bee hives very near to the field to save bee's energy
- Place bee hives in field at 10% flowering of crop
- Place2- 3 colonies per ha
- The colonies should have full strength of bees
- Allow sufficient space for pollen and honey storage
- Provide artificial sugar syrup and water if required

Scavengers

These are insects which feed upon the dead and decaying plant and animal matter. Since insects help to remove from the earth surface the dead and decomposing bodies, which would otherwise be a health hazard, they are referred to as scavengers.In addition to cleaning the filth from human habitations, these insects help to convert those bodies into simpler substances before recycling them back to soil, where they become easily available as food for growing plants. In this respect termite, maggots of many flies and larvae and adults of beetles are important.

Red Ants for control of tea mosquito

Red ant colonies are established in cashew garden by bringing the colonies from other trees to cashew and connecting the trees with jute or plastic ropes to facilitate ant movement. Pieces of meat or fish should be kept in the other end of rope in the tree which needs to be invaded. Ant colonies should be kept in the evening time and provide food in the form of pieces of meat or fish for successful colonization.

Rearing of red ants

Under the trunk of tree apply 5 baskets of ash has to be placed near the tree collar region. Clay soil or loamy soil taken from tank. The soil should not be easily dissolved when moist. Both are mixed together. Above this sprinkle a layer of jaggery powder. This will attract red ants and make nesting nearby. The increased ant population will be beneficial in controlling stem borers or any caterpillar pests in the garden.

Wild birds

It has been shown that a pair of blue tits can consume 10,000 caterpillars and one million aphids in a 12 month period. The installation of tit boxes is a worthwhile activity. Wrens, thrushes and blackbirds similarly contribute to the control of garden insects.

The black-kneed capsid

The black-kneed capsid (*Blapharidopterus angulatus*) is an insect found on fruit trees alongside its pestilent relative, the common green capsid. It eats more than 1000 fruit tree red spider mites per year. Its eggs are laid in August and survive the winter. Winter washes used by professional horticulturalists against apple pests and diseases often kill off this useful insect. The closely related anthocorid bugs, such as *Anthocoris nemorum*, are predators on a wide range of pests, such as aphids, thrips, caterpillars and mites, and have recently been used for biological control in greenhouses.

Lacewings

Lacewings, such as *Chrysopa carnea*, lay several hundred eggs per year on the end of fine stalks, located on leaves. Several are useful horticultural predators, their hairy larvae eating aphids and mite pests, often reaching the prey in leaf folds where ladybirds cannot reach.

Ladybird beetle

Almost 40 species of lady bird beetle are predatory among pest in horticultural crops. The red two-spot ladybird (*Adalia bipunctata*) emerges from the soil in spring, mates and lays about 1000 elongated yellow eggs on the leaves of a range of weeds, such as nettles, and crops such as beans, throughout the growing season. Both the emerging slate-grey and yellow larvae and the adults feed on a range of aphid species. Wooden ladybird shelters and towers are now available to encourage the overwintering of these useful predators.

A worrying development in the last few years has been the rapid spread and increase of the harlequin ladybird from South-East Asia. This species is larger (6–8 mm long) and rounder than the two-spot species (4–5 mm). It has a wider food range than other ladybird species, consuming other ladybird's eggs and larvae, and eggs and caterpillars of moths. Furthermore, it is able to bite humans and be a nuisance in houses when it comes out of hibernation.

Carabid beetles

The ground beetle (such as *Bembidion lampros*), a 2 cm long black species, is one of many active carabid beetles that actively predates on soil pests such as root fly eggs, greatly reducing their numbers.

Mites and spiders

Predatory mites such as *Typhlodromus pyri* eat fruit tree red spider mite and contribute importantly to its control. The numerous species of webforming and hunting spiders help in a very important but unspecific way in the reduction of all forms of insects.

Wasps

The much maligned **common wasp** (*Vespula vulgaris*) is a voracious spring and summer predator on caterpillars, which are fed in a paralyzed state to the developing wasp grubs. **Digger wasps** also help control caterpillar numbers and benefit from dead hollow stems of garden plant which they use as nests all year round.

Microbial control

Microbial control refers to the exploitation of disease causing organism to reduce the population of insect pest below the damaging levels. Steinhaus (1949) Coined the term 'Microbial Control' when microbial organisms or other products (toxins) are employed by man for the control of pests on plants, animals or man.

1. **Bacteria :** More than 100 pathogenic bacteria were recorded of which i. *Bacillus thuringiensis* (*B.t.*) is important and is isolated from flour moth, *Ephestia kuehniella* by Berliner (1915) *B.t.* known as a bacterial insecticide is now being using by farmers mostly on lepidopterous larvae. It can infect more than 150 species of insects. The entry of the bacteria is by ingestion of the bacteria, which infect the midgut epithelial cells and enter the haemolymph to sporulate and cause septicemia.

Properties of *Bt*

1. Highly pathogenic to lepidopterous larvae
2. Non-toxic to man
3. Non-phytotoxic
4. Safer to beneficial insects
5. Compatible with number of insecticides
6. So far no resistance is developed in insects

7. Synergistic in combination with certain insecticides like carbaryl
8. Available in different formulations (Trade names Thuriocide, Delfin, Bakthane,Biobit, Halt, Dipel etc).
9. Formulation is so standardized that 1 gm of concentration spore dust contains 100 million spores *Bacillus popilliae* (available as Doom) causes milky disease on Japanese beetle, *Popillia japonica*

2. Viruses: NPV (Borrellina virus): About 300 isolates of Nuclear polyhedral virus have been isolated from the order Lepidoptera. Among these viruses Baculoviruses (Baculoviridae) are successful in IPM. The NPV is observed to affect 200 species of insects like *Corcyra cephalonica, Pericallia ricini, Amsacta albistriga, Spodoptea litura, Heliothis armigera* etc., by ingestion. The virus infected dead larvae hanging upsidedown from plant parts (Tree top disease). The cuticle becomes fragile, rupturing easily when touched, discharges liquefied body fluids. NPV multiplies in insect body wall, trachea, fat bodies and blood cells. The polyhedra are seen in nuclei. The polyhedral bodies enlarge in size destroying the host nuclei to get released into the insect bodycavity.

Fungi: The fungal disease occurrence in insects is commonly called as mycosis. Most of the entomopathogenic fungi infect the host through the cuticle. The process of pathogenesis begins with

- Adhesion of fungal infective units or conidia to the insect epicuticle
- Germination of infective units on cuticle
- Penetration of the cuticle
- Multiplication in the haemolymph
- Death of the host (Nutritional deficiency , destruction of tissues and releasing toxins)
- Mycelial growth with invasion of all host organs
- Penetration of hyphae from the interior through the cuticle to exterior of the insect
- Production of infective conidia on the exterior of the insect.

Most of the entomopathogenic fungi infect their hosts by penetration of the cuticle by producing cuticle digesting enzymes (Proteases , lipases chitinases).The typical symptoms of fungal infection are, mummified body of insects and it does not disintegrate in water and body covered with filamentous mycelium.More than 5000 species of entomopathogenic fungi are recorded. Important species are, *Entomophthora, Metarhizium, Beauveria, Nomuraea* and *Verticillium.*

7. Farmers wisdom in pest management

The knowledge of traditional agriculture with millions of farmers should be utilized and modern technology in agriculture should be blended with traditional wisdom. The following are certain practices of farmers which they have been following time immemorial

a. Diluted cow dung slurry sprinkled to hasten paddy germination.
b. Coconut fronds cut into small bits erected as perches in field to attract nocturnal birds which preys upon rats.
c. Chilli mash and garlic juice sprayed to control rice earhead bug.
d. Application of common salt at 1 - 1.5 kg/ palm of coconut gives insect resistance and prevents button shedding.
e. Use of scarecrows to ward off bird pests in day time, which also serve as perches to nocturnal predatory birds.
f. Use of Kavankal where stones are released from slings to scare birds
g. Ploughing of field during Agninakshatra (April-May) when temperature is around 40 – 45° C brings about killing of soil insects, nematodes and pupae of lepidopteran pests.
h. Treating stored pulses with red earth (1 kg of red earth: 1 kg of pulses) to prevent insect damage.
i. Use of Tanjore bow trap, a common traditional gadget to kill rats in rice fields of Cauvery delta.
j. 'Vrikshayurveda", a science of plant health, similar to 'Ayurveda" which is science of human life deals with maintenance of plant health and provides literature on control measures for control of pests and diseases.

Table 13. Farm level practice for insect pest control

Cropping Techniques	Pest Checked
Ploughing	Red hairy caterpillar (*Amsacta albistriga*)
Puddling	Rice mealy bug (*Brevennia rehi*)
Trimming and plastering	Rice grasshopper (*Hieroglyphus daganensis*)
Pest free seed material	Potato tuber moth (*Phthorimaea operculella*)
High seed rate	Sorghum shoot fly (*Atherigona soccata rondani*)
Rogue space planting	Rice brown planthopper (*Nilaparvata lugens*)
Plant density	Rice brown planthopper (*Nilaparvata lugens*)
Earthing up	Sugarcane whitefly (*Aleurolobus barodensis*)
Detrashing	Sugarcane whitefly (*Aleurolobus barodensis*)
Destruction of weed hosts	Citrus fruit sucking moth (*Othreis fullonica*)
Destruction of alternate host	Cotton whitefly (*Bemisia tabaci*)
Flooding	Rice armyworm (*Pseudaletia unipuncta*)
Trash mulching	Sugarcane early shoot borer (*Chilo infuscatellus*)
Pruning / topping	Rice stem borer (*Scirpophaga incertulas*)
Intercropping	Sorghum stem borer (*Chilo partellus*)
Trap cropping	Diamond back moth (*Plutella xylostella*)
Water management	Brown planthopper (*Nilaparvata lugens*)
Timely harvesting	Sweet potato weevil (*Cylas formicarius*)

Community level practices

- Synchronized sowing : Dilution of pest infestation (eg) Rice, Cotton
- Crop rotation : Breaks insect life cycle
- Crop sanitation
  a) Destruction of insect infested parts (eg.) Mealy bug in brinjal
  b) Removal of fallen plant parts (eg.) Cotton squares
  c) Crop residue destruction (eg.) Cotton stem weevil

8. Use of plant products / botanicals

Neem leaf extract

- 1 kg of green neem leaf soaked overnight in 5 litres of water, crushed and the extract is filtered. Add 10 ml of emulsifier (neutral pH adjuvant)
- This is beneficial against leaf eating caterpillars, grubs, locusts and grasshoppers.

Neem cake extract

- 100 g of neem cake is soaked overnight in 1 litre of water in a muslin pouch, crushed and the extract is filtered. Add 1 ml of emulsifier (neutral pH adjuvant) per llitre of water

Neem oil spray

- 15-30 ml neem oil is added to 1 litre of water and stirred well. To this emulsifier is added (1ml/1 litre).

Use of Cycas flowers

The flower of Cycas are cut into pieces, wrapped in straw and placed in the field @ 25–30 pieces /hectare. The odour that is emitted from this flower prevents the entry of earhead bugs for two weeks. By this time, the milky stage is over and the grain matures without any interruption.

Five leaf extract

This extract is prepared using the leaves of five different plants. Leaves with the characteristics described below can be used for the purpose:

In a mud pot, any five of the below mentioned plant leaves pound well

Plants with milky latex – e.g., *calotropis, nerium, cactus* and *jatropha.*
Plants which are bitter – e.g., *neem, andrographis, tinospora* and *leucas.*
Plants that are generally avoided by cattle – e.g., *adathoda, Ipomea fistulosa.*
1kg each
Aromatic plants – e.g., *vitex, ocimum.*
Plants that are not affected by pests & diseases – e.g., *morinda, Ipomea fistulosa.*

Then add

Water - 10 litre Tie the mouth of the pot tightly with a cloth

Cow's urine - 1 litre

Asafoetida - 100 g Mixed well daily for week

- It can be used after a week after filtration. Cow urine is used for disease control and asafoetida prevents flower dropping, enhancing the yield.

Jatropha leaf extract

In a mud pot, pound

jatropha leaves - 12.5 kg

To this add

Water - 12.5 litres

- Allow to ferment for 3–7 days.
- Filter and use the extract for spraying (after diluting with 10 parts of water) for one hectare area.

Turmeric rhizome extract

- Shred one kilo of turmeric rhizomes. To this, add four litres of cow urine, mix well and filter. Dilute with 15–20 litres of water.
- For every litre of the mixture, add 4 ml of khadi soap solution. This helps the extract stick well to the surface of the plant.

Cow dung extract

- Mix one kilo of cow dung with ten litres of water and filter using a gunny cloth.
- Dilute the solution with five litres of water and filter again the result can be used for spraying.

Fermented curd water

In some parts of central India fermented curd water (butter milk or *Chaach*) is also being used for the management of white fly, jassids aphids, etc.

Take the distasteful butter milk and add equal amount of water in it. Keep it for two days in a semi-shaded place. Now take it and add 40 lit of water in 10 lit of the extract to form 50 lit of the solution. Spray this solution in 1 hectare of field in such a way that all plants get bath in the fogging spray in the early morning time.

Herbal Pesticide formulation

- 500g neem seeds, 1000g tobacco, 100g *Acorus calamus*, 250g As a ofoetida and 50g *Sapindus emarginatus* seeds are ground and the extract is sprayed for one acre cotton to control pests.

Neem-cow urine extract

- 5 kg of neem leaves, 5 lit of cow urine, 2 kg of cow dung, 100 lit of water.
- Crush all ingredients and ferment for 24 hours with intermittent stirring, filter and squeeze the extract and dilute to 100 lit of water.
- Use this extract to fill in the spray machine and spray it over one acre of the crop.

PUSH-PULL strategy for insect pest management

The 'push-pull' strategy, a novel tool for integrated pest management programs, uses a combination of behavior-modifying stimuli to manipulate the distribution and abundance of insect pests and/or natural enemies. In this strategy, the pests are repelled or deterred away from the main crop (push) by using stimuli that mask host apparency or are repellent or deterrent. The pests are simultaneously attracted (pull), using highly apparent and attractive stimuli, to other areas such as traps or trap crops where they are concentrated, facilitating their control.

The development of a reliable, robust, and sustainable push-pull strategy requires a clear scientific understanding of the pest's biology and the behavioral/chemical ecology of the interactions with its hosts, conspecifics, and natural enemies. The specific combination of components differs in each strategy according to the pest to be controlled (its specificity, sensory abilities, and mobility) and the resource targeted for protection.

Among several push-pull strategies under development or used in practice for insect pest control, the most successful example of the push-pull strategy currently being used by farmers was developed in Africa for controlling stem borers on cereal crops.

The push-pull strategy for cereal stemborers involves trapping stemborers on highly attractant trap plants (pull) while driving them away from the main crop using repellent intercrops (push). Plants that have been identified as effective in the push-pull tactics include Napier grass (*Pennisetum purpureum*), Sudan grass (*Sorghum vulgare sudanense*), molasses grass (*Melinis minutiflora*), and desmodium (*Desmodium uncinatum* and *Desmodium intortum*).

How push-pull strategy works

The push-pull strategy undertakes a holistic approach in exploiting chemical ecology and agrobiodiversity. The plant chemistry responsible for stem borer control involves release of attractive volatiles from the trap plants and repellent volatiles from the intercrops.

Benefits of push-pull strategy

The principles of the push-pull strategy are to maximize control efficacy, efficiency, sustainability, and outputs, while minimizing negative environmental effects. Cultivation of Napier grass for livestock fodder and soil conservation now assumes an additional rationale as a trap plant for stemborer management. Similarly, desmodium, a nitrogen-fixing legume, already grown for improving soil fertility and for quality fodder, is also an effective stem borer repellent and striga weed suppressant. Intercropping desmodium with maize reduces the need for external mineral nitrogen inputs, which are costly and unaffordable by most small-holder rural people, and improves the use efficiency of other inputs.

Disease management

Organic farming involves a comprehensive approach to soil health, crop health, and the maintenance of a biologically dynamic and diverse agro-ecosystem. Many organic farming practices favorably impact plant health. Examples of these practices are crop rotation and building soil organic matter, through cover crop amendments and inclusion of compost. In fact, organic growers often observe a reduction in some diseases following the transition from conventional to organic methods. The plant health benefits conferred by organic production practices can be further enhanced through additional cultural and biological practices that suppress disease.

What Organisms Causes Disease?

1. Fungi
2. Bacteria
3. Viruses
4. Nematodes

What types of symptoms can you expect from these organisms?

1. **Fungi** cause the great majority, estimated at two-thirds, of infectious plant diseases. They include all spots, lesions, blights, yellowing of leaves, wilts, cankers, rots, fruiting bodies, mildews, molds, leaf spots, root rots, cankers, and blotches. Fungi are typically spread by wind, rain, soil, mechanical means and infected plant material.
2. **Bacteria** Bacteria cause any of the four following main problems. Some bacteria produce enzymes that breakdown the cell walls of plants anywhere in the plant. This causes parts of the plant to start rotting (known as 'rot'). Some bacteria produce toxins that are generally damaging to plant tissues, usually causing early death of the plant. Others produce large amounts of very sticky sugars; as they travel through the plant, they block the narrow channels preventing water getting from the plant roots up to the shoots and leaves, again causing rapid death of the plant. Finally, other bacteria produce proteins that mimic plant hormones.These lead to overgrowth of plant tissue and form tumours.
3. **Viruses** mostly cause systemic diseases. Generally, leaves show chlorosis or change in colour of leaves and other green parts. Light green or yellow patches of various shades, shapes and

sizes appear in affected leaves. These patches mayform characteristic mosaic patterns, resulting in general reduction in growth and vigour of the plant.They include mottling, leaf and stem distortions, mosaic patterns, rings and stunting. Viruses cause interesting symptoms, some are beautiful. Viruses are spread by mechanical means, vectors and in plant material.

4. **Nematodes** cause wilting, stunting, yellowing of entire plants. This is because the roots of the plant are infected and the plant is starving or thirsty. Nematodes are spread by soil on equipment or workers boots or on infected plant material.

Principles of Plant Disease Control

1. **Avoidance**: prevents disease by selecting a time of the year or a site where there is no inoculum or where the environment is not favorable for infection.
2. **Exclusion**: prevents the introduction of inoculum.
3. **Eradication**: eliminates, destroy, or inactivate the inoculum.
4. **Protection**: prevents infection by means of a toxicant or some other barrier to infection.
5. **Resistance**: utilizes cultivars that are resistant to or tolerant of infection.
6. **Therapy**: cure plants that are already infected

1. Preventive practices

1) *Usage of resistant varieties*

- Choose varieties which are well adapted to the local environmental conditions (temperature, nutrient supply, pests and disease pressure), as it allows them to grow healthy and makes them stronger against infections of pests and diseases.

2) *Use of suitable cropping systems*

- Mixed cropping systems: can limit pest and disease pressure as the pest has less host plants to feed on and more beneficial insect life in a diverse system.
- Crop rotation: reduces the chances of soil borne diseases and increases soil fertility.

3) *Use of balanced nutrient management*

- Moderate nutrition: steady growth makes a plant less vulnerable to infection. Too much fertilization may result in salt damage to roots, opening the way for secondary infections.
- Balanced potassium supply through organic sources wood ash comosted sdurdust contributes to the prevention of fungi and bacterial infections

4) *Input of organic matter*

- Increases micro-organism density and activity in the soil, thus decreasing population densities of pathogenic and soil borne fungi.

5) *Application of suitable soil cultivation methods*

- Facilitates the decomposition of infected plant parts.
- Regulates weeds which serve as hosts for diseases.
- Protects the microorganisms which regulate soil borne diseases.

6) *Use of good water management:*

- No water logging: causes stress to the plant, which encourages pathogens infections.
- Avoid water on the foliage, as water borne disease spread with droplets and fungal disease germinate in water.

7) *Selection of optimum planting time and spacing:*

- Most pests or diseases attack the plant only in a certain life stage; therefore it's crucial that this vulnerable life stage doesn't correspond with the period of high pest density and thus that the optimal planting time is chosen.
- Sufficient distance between the plants reduces the spread of a disease.
- Good aeration of the plants allows leaves to dry off faster, which hinders pathogen development and infection.

8) *Use of proper sanitation measures*

- Remove infected plant parts (leaves, fruits) from the ground to prevent the disease from spreading.

- Eliminate residues of infected plants after harvesting.
- Incorporating the residue into the soil hastens the destruction of disease pathogens by beneficial fungi and bacteria. Burying diseased plant material in this manner also reduces the movement of spores by wind.

9) *Forecasting*

Producers should pay attention to the forecasts for various pest and disease infestations for each crop year. Maps of these forecasts are usually available for many of the major destructive insects such as grasshoppers and wheat midge, as well as some diseases. Agro meteorological warning and forecast can help in this way.

2. Cultural Control

Cultural control is your first line of defense.

A healthy crop

By giving plants the right growing conditions they will be more able to resist pests and diseases. Also, the right choice of crop will help to deter pests and disease. A crop growing in an area where it is not suited is more likely to be attacked. You should take account of the soil type, the climate, the altitude, the available nutrients and the amount of water needed when selecting your crops. Plants will only, yield well and resist pests and diseases if they are grown under the most suitable conditions for that particular plant. To help ensure a healthy crop, weeding should be done early and regularly to stop weeds from taking nutrients which should be going to the crop.

Choice of tolerant or resistant crops and varieties

The use of crops and varieties tolerant or even resistant against common diseases is an effective measure to lower risks of pest and disease damage.In organic farming, the selection of varieties with partial resistance or field tolerance to disease is practical and even preferable to high-level resistance. There are more commercial varieties with disease resistance than are known for pest resistance. Therefore, for disease resistance, the local knowledge of farmers and advisers about the characteristics of traditional and local crop varieties is of highvalue. Even 'resistant' varieties need to be adapted under local climatic conditions for effective resistance.

Field hygiene and sanitation

Promote healthy soils and healthy plants. Healthy soil is the hallmark of organic agriculture. An unhealthy plant is very attractive to diseases. The use of disease free seeds and planting materials is a very effective tool to prevent development of seed-borne pests and diseases. Certified seeds are normally clean, but if such seeds are not available to the farmers, the seeds should be treated before use to eliminate seed-borne diseases.

Sanitation of existing crops, especially perennial crops should be done regularly. Poorly managed or abandoned perennial crops can result in build upof pest and disease problems. All damaged plant materials and rotten fruits from the ground must be either burned or deeply buried at least 50 cm deep. Pruning eliminates inoculum in perennial crops. All infected branches orshoots should be cut at least 20 cm below the visible damage. Pruning also improves aeration and light exposure to the crown, which contribute to prevention of diseases.

Eradication of alternate and collateral hosts

Eradication of alternate hosts will help in management of many plant diseases.

- The macrocyclic rusts needs a alternate hosts to complete its life cycle and if the alternate host is eradicated the pathogen can be checked.
- E.g. *Puccinia graminis tritici* alternate host barberry can be eradicated to break the disease cycle.
- The pathogen survives in the **collateral hosts** during the off season and hence the collateral hosts can be eradicated to check the primary inoculum.
- Eg.The collateral host of blast viz; *Panicum repens, Digitaris marginata, Echinochloa crusgali* can be eradicated to manage the disease.

Choice of geographic area

The elements of a site which will affect disease are numerous, and they interact with the particular crop and management strategy being considered.

Some diseases are serious in wet areas e.g. Smut and ergot of bajra, karnal bunt of wheat. *Phytophthora* blight of colocasia is serious during rainy season.

Dry areas

Fusarium oxysporum f.sp. *lycopersici* is serious in dry belts.

Disease History

Many sites simply cannot support particular crops, at least for a time, because of the persistence of a disease which may afflict that species. Cole crops must not be planted into soil which harbored clubroot of crucifers in the past; in other cases a suitable rotation will sanitize the soil after a time, or lower initial inoculum levels below the economic threshold.

Soil Drainage

Many diseases, particularly root rots and damping off caused by fungi like *Pythium* spp., are favored by heavy, poorly-drained soils. Some of these soils with a history of infestation simply cannot be used to grow most crops without installing drainage tile. This is one of the many elements of soil management that affects disease.

Selection of field

- The **sick soil** in which the soil borne inoculum persists should not be selected for the ensuing cropping season.
- E.g. Red rot of sugarcane *Collectrichum falcatum* persists in the soil for few months.
- Panama wilt of banana-*Fusarium oxysporum* f.sp.*cubense*
- Bacterial wilt of potato-*Pseudomonas solanacearum*
- Smut of bajra -*Tolyposporium pencillariae*
- Ear cockle of wheat-*Corynebacterium tritici* and *Anguina tritici*

Crop rotation

Continuous cultivation of the same crop in the same field helps in the perpetuation of the pathogen in the soil. Soils which are saturated by the pathogen are often referred as sick soils. To reduce the incidence and severity of many soil borne diseases, crop rotation is adopted. Crop rotation is applicable to only root inhabitants and facultative saprophytes, and may not work with soil inhabitants.Ex: Panama wilt of banana (long crop rotation), wheat soil borne mosaic (6 yrs) and clubroot of cabbage (6-10 yrs), etc.,

Manures

The deficiency or excess of a nutrient may predispose a plant to some diseases. Excessive nitrogen application aggravates diseases like stem rot, bacterial leaf blight and blast of rice. Nitrate form of nitrogen increases many diseases,whereas, phosphorus and potash application increases the resistance of the host. Addition of farm yard manure or organic manures such as green manure, 60-100 t/ha,helps to manage the diseases like cotton wilt, Ganoderma root rot of citrus, coconut, etc.

Mixed cropping

Root rot of cotton (*Phymatotrichum omnivorum*) is reduced when cotton is grown along with sorghum. Intercropping sorghum in cluster bean reduces the incidence of root rot and wilt (*Rhizoctonia solani*).

Soil amendments

Application of organic amendments like saw dust, straw, oil cake,etc., will effectively manage the diseases caused by *Pythium, Phytophthora, Verticillium,Macrophomina, Phymatotrichum* and *Aphanomyces*. Beneficial micro-organisms increases in soil and helps in suppression of pathogenic microbes.

Plant spacing

Close spacing raises atmospheric humidity and favours sporulation by many pathogenic fungi. A spacing of 8'x 8' instead of 7'x7' reduces sigatoka disease of banana due to better ventilation and reduced humidity. High density planting in chillies leads to high incidence of damping off in nurseries.

- Overcrowding results in Damping off of seedlings, late blight of potato, downy mildew of grapevines -spread fast in close planting.Bacterial diseases

Irrigation

The amount, frequency and method of irrigation may affect the dissemination of certain plant pathogens. Many pathogens, including, *Pseudomonas solanacearum, X. campestris pv. Oryzae* and *Colletotrichum falcatum* are readily disseminated through irrigation water. High soil moisture favours root knot and other nematodes and the root rots caused by species of *Sclerotium, Rhizoctonia, Pythium, Phytophthora, Phymatotrichum*, etc. Flooding destroys *Pseudomonas solanacearum, Verticillium dahliae, Fusarium oxysporum* f sp *cubense* and the nematodes *Radopholus similis*.

Amount of irrigation

- In some cases insufficient water causes a plant to become more susceptible to disease.
- Charcoal rot of sorghum and root rot of cotton induced by *Macrophomina phaseolina* are more severe when plants are subjected to water stress.
- Irrigating from infected to healthy fields should be avoided
- Having a trench to minimise spread

Timing

Exactly when planting or transplanting is done can be used to avoid disease. A grower must be familiar with the optimal conditions for disease development and the local climate to take advantage of this option.

- Generally, seed should be planted when temperature and moisture are optimum for rapid germination and growth. Seed rot and damping off are more probable, especially for untreated organic seed, under wet and cool conditions where germination and growth are delayed.

- Late planting can help control barley yellow dwarf in rice, Fusarium root rots of beans, downy mildew of sunflower, and stalk rots of corn. In Florida, late planting of winter tomatoes (end of November) means cooler weather, and this helps control Fusarium wilt.
- Early planting can help reduce diseases such as downy mildew and leaf blights of corn, powdery mildew of peas, and common smut and fungal root rots of wheat. Note that conflicts may arise and planting time must be adjusted to a grower's particular location and disease pressures. Fusarium root rot of peas is favoured by early planting, but powdery mildew may be controlled by early planting.

Soil pH

The pH (acidity or alkalinity) of a soil can affect some diseases. Changing the pH can reduce the problem.

For example, potato scab is less severe in more acid (pH below 7) soils. A layer of grass mowings added to the bottom of the potato trenches at planting time will make the soil more acid and reduce the disease.

Clubroot is less severe in alkaline conditions (pH above 7) therefore liming the soil to make it more alkaline can reduce the problem.

Rouging

Removal of diseased plants or their affected organs from field, which prevent the dissemination of plant pathogens.

- Rouging of loose smut of wheat, barley, sorghum, maize, red rot of sugar cane, yellow mosaic virus infected plants are done to eradicate the source of inoculum.
- This procedure is practical only when size of the plot is small and number of diseased plant is less.

Compost application

Soils rich in organic matter are shown to increase soil biodiversity and help to create and abundance of beneficial soil microorganisms.

Using compost has been shown to increase the suppressiveness of the soil by encouraging beneficial microorganisms, as well as inducing disease resistance in plants by simply having healthier plants.

3. Physical methods

Hot water treatment

Hot water treatment of own seed to prevent seed-borne diseases such as blackrot, black leg, black spot and ring spot of crucifers is very effective. It reduces the seed-borne pathogens such as *Alternaria* spp., *Colletotrichum* spp., *Phoma* spp., *Septoria* spp. and bacterial pathogens (*Pseudomonas* spp. and *Xanthomonas* spp). However, hot water treatments are delicate as seeds can rapidly be destroyed by too hot temperatures.Therefore, specified temperature and time intervals must be strictly followed in order to maintain seed viability.

- Potato tuber, banana suckers: 55° C for 10 minutes
- Spinach, brussels sprouts, cabbage, pepper, tomato, eggplant: 50° C for 30 minutes
- Broccoli, cauliflower, carrot, collard, kale, kohlrabi, turnip: 50°C for 20 minutes
- Mustard, cress, radish: 50°C for 15 minutes
- Lettuce, celery, celeriac: 47°C for 30 minutes

Soil solarization: Soil solarization or slow soil pasteurization is the hydro/thermal soil heating accomplished by covering moist soil with polyethylene sheets as soil mulch during summer months for 4-6 weeks. Soil solarization was developed for the first time in Israel (Egley and Katan) for the management of plant pathogenic pests, diseases and weeds.

Mulching

Material covering the ground on which your plants are growing can help with disease, but in some cases encourage it. Organic mulches can promote beneficial microbes and encourage disease-suppressive soils. They can also hinder dispersal of rain-splashed inoculum, while plastic mulches can exacerbate splash dispersal.

Timing is important. Early in the season, mulch can encourage damping off by keeping soil from warming and drying, especially problematic for disease-susceptible young stems. But later in the year, fruits such as tomatoes or cucurbits can be kept out of contact with soil and avoid rot thanks to intervening with mulch.

Soil sterilization: Soil can be sterilized in green houses and sometimes in seed beds by aerated steam or hot water. At about 50°C, nematodes, some oomycetous fungi and other water molds are killed. At about 60 and 72°C, most of the plant pathogenic fungi and bacteria are killed. At about 82°C, most weeds, plant pathogenic bacteria and insects are killed. Heat tolerant weed seeds and some plant viruses, such as TMV are killed at or near the boiling point (95-100°C).

4. ITK in disease control

- A mixture of ash (2-3 kg) and 1 liter of castor oil is spread on a sedd bed of a size of about 100 m^2. The application is repeated 2-3 times at intervals of 7-10 days. This protects against soil borne diseases in tobacco nurseries.
- A mixture of 2 kg turmeric powder and 8 kg wood ash is used as dust over leaves for treatment against powdery mildew.
- Ginger powder at 20 g/lit of water and sprayed thrice at interval of 15 days can be used for powdery mildew and other fungal diseases.
- Handful of slaked lime applied at the base of tomato plant can combat damping off disease.
- Cattle and goat urine have fungicidal properties. 2 cups of cattle urine with 5 ml peppermint oil and 10 lit of water can be used to control fungal diseases on grapes.

Weed management in organic farms

Weeds are the plants, which grow where they are not wanted (Jethro Tull, 1731). Weeds can also be referred to as plants out of place. All plants that compete with our crops for nutrients, water and light, and reduce their harvest and quality are considered weeds. Some plants may also poison domesticated animals when growing on pasture land.

On the other hand weeds provide cover to the soil helping to reduce erosion. They also contribute to biological diversity in the crop fields by providing good living conditions and food for beneficial organisms. Thus we should not struggle to completely eradicate weeds, as they play an important role in the farm ecosystem.

Of all the production challenges facing organic growers, weed management remains for most one of the most difficult, frustrating, expensive, and time- consuming management aspects of producing a crop for market.

Organic farmers do not try to eradicate all weeds. Rather, they try to keep weed populations in and around the fields at such low levels that they do not become dominant. Poisonous and destructive weeds like Striga, curcutta etc., however, should be completely removed and destroyed.

What Weeds are doing in my field beyond competition?

Weeds are nature's way of covering soil that has become exposed by fire, flood, landslide, windstorms, clear-cutting, clean tillage, herbicides, overgrazing, or other disturbance. Bare soil is hungry and at risk. The soil life, so vital to soil fertility, goes hungry because the normal influx of nourishing organic compounds from living plant roots has been cut off for the time being. The exposed soil surface is at risk of erosion by rain or wind, especially if root systems have also been removed or disrupted. Pioneer plants – weeds – are those species that can rapidly cover bare soil and begin performing a number of vital ecological functions:

- Protect the soil from erosion.
- Stabilizes the soil
- Nectar for bees
- Human consumption as leafy vegetables
- Replenish organic matter, feed and restore soil life.
- Absorb, conserve, and recycle soluble nutrients that would otherwise leach away.
- Absorb carbon dioxide from the atmosphere.
- Restore biodiversity.

- Economic Importance (Nutsedge – For making insence sticks)
- Reclamation of Alkali Soils (*Argemone Mexicana*) @ 2.5 tonnes/ha is useful for reclamation of alkali soils.)
- Provide habitat for insects and animals.
- Weeds are sources of pesticides e.g *Chrysanthemum cinerariifolium* which provides insecticide pyrethrum.
- Weeds provide food and cover for animal. Wildlife generally depends on weeds for survival as food and shelter.
- Act as a medicinal source for pharmaceutical industries and used in ethnomedicines.

Weeds are an indicator of a soil's health

Farmers constantly battle with the weeds, but weeds can have a useful purpose - they can be used as soil indicators. Simply by observing the most prevalent weeds that are growing in a specific area, they can indicate if the soil is acidic or alkaline, whether the soil is a healthy, balanced or if it's depleted. Weeds can indicate a poorly draining soil, or a soil that is unable to retain moisture. Weeds can even indicate if the soil is unbalanced, being overly rich in one nutrient and deficient in others.

When using weeds as a soil indicator, observe several of the most prevalent types of weeds to get an accurate soil assessment. If on cultivated field suddenly appears certain weed species, i.e. if this type is dominating, it indicates mistakes in cultivation technology management and soil condition. Weeds can be a headache, but they can also be very helpful if we know a few basic principles. Weeds give us clues to the health of our soil.

1. Weed Management during Transition

Organic vegetables are often established on old hay fields or pastures to shorten time to certification. These fields may have severe infestations of perennial weeds and dense seed banks of annuals.

- Avoid planting vegetables and cash crops the first year.
- Start with a cover crop.

- Till in the cover crop before perennials get large or annuals go to seed.
- Repeat at 4- to 6-week intervals all summer.
- Tilled fallow will deplete the seed bank and exhaust perennial roots.
- If weeds are still likely, start with short season crops like pulses that will be out before weeds go to seed.

2. Preventive methods: Preventing or denying establishment or entry of a new weed species into an area

- Avoid using crop that are infested with weed seeds for sowing
- Avoid feeding screenings and other material containing weed seeds to the farm animals.
- Avoid adding weeds to the manure pits.
- Clean the farm machinery thoroughly before moving it from one field to another. This is particularly important for seed drills
- Avoid the use of gravel sand and soil from weed-infested
- Inspect nursery stock for the presence of weed seedlings, tubers, rhizomes, etc.
- Keep irrigation channels, fence-lines, and un-cropped areas clean
- Use vigilance. Inspect your farm frequently for any strange looking weed seedlings. Destroy such patches of a new weed by digging deep and burning the weed along with its roots. Sterilize the spot with suitable chemical.
- Quarantine regulations are available in almost all countries to deny the entry of weed seeds and other propagules into a country through airports and shipyards.

3. Cultural or Agronomic methods

Field Preparation

- The field has to be kept weed free. Flowering of weeds should not be allowed. This helps in prevention of build up of weed seed population in the fields.

- Irrigation channels are the important sources of spreading weed seeds. It is essential, therefore, to keep irrigation channels clean.
- Deep ploughing in summer exposes underground parts like rhizomes and tubers of perennial and obnoxious weeds to scorching summer sun and kills them.
- Conventional tillage which includes 2 to 3 ploughings followed by harrowing decreases the weed problem.
- Running blade harrows cuts weeds and kills them.
- In lowland rice, puddling operation incorporates all the weeds in the soil which would decompose in course of time.

Pasturing

In perennial crops like coffee, mangoes, avocadoes or cocoa, the use of sheep and goats to reduce rampant weed growth is becoming common. In case of cattle, broadleaf weeds tend to predominate due to the cattle preference for grasses. Therefore, it is necessary to rotate with sheep and goats which prefer broadleaves to overcome this selective grazing.

- Prevent dissemination of weeds by eliminating them before seed dispersal.
- Avoiding the introduction of weed seeds into the fields through tools or animals; and by using only weed free seed material.

Sowing time and density

Optimum growing conditions enhance the optimum crop plant development and their ability to compete against weeds. Proper crop spacing will ensure that minimum space is available for the growth of weeds and will minimize competition with weeds.This will effectively restrict weed development. In order to apply this approach, the limiting weeds must be known and the seasons in which they occur. A weed calendar of the area or region, if available, might be of help. It will be used to manage weeds in a targeted fashion with proper timing and effect.

Planting in narrow rows

Keeping all other factors constant, narrow spaced crops smother weeds more efficiently than wide spaced crop. Wide spaced crop allows more light to reach weeds. Hence, weed growth is more.

Planting direction

Crops sown in North-South direction suppress weed growth better than East-West direction. East-West crop allows sunlight to reach weeds throughout the day, whereas, North-South sown crop utilizes more light, shades weeds and does not allow light to reach weeds.

Plant density

In the crop-weed competitive situation, higher number of plants gain, comparative edge, more plant result in broader canopy cover and thus higher weed suppression capacity

- Proper time of planting
- Crop rotation

Weeds get familiarized with canopy cover, growth habit and cultivation practices adopted for a particular crop. Weed management is difficult in cropping systems where the same crop is grown year after year. On the other hand with a change in cropping sequence, cultivation practices vary and so also the canopy cover and growth habit of the crop

- Inter cropping

Planting Method

1. Sowing of clean crop seeds without weed seeds should be done. It is a preventive method against introduction of weeds.
2. Sowings are taken up one to three days after rainfall or irrigation depending on soil type. Weeds already present in the soil start germinating within two or three days.
3. Sowing operation with seed drill removes some of the germinating weeds.
4. Transplanting is another operation which reduces weed population. Since, the crop has an additional advantage due to its age.

Stale seed bed

A stale seedbed is one where initial one or two flushes of weeds are destroyed before planting of a crop. This is achieved by soaking a well prepared field with either irrigation or rain and allowing the weeds to germinate. At this stage a shallow tillage or non- residual herbicide like paraquat may be used to destroy the dense flush of young weed seedlings. This may be followed immediately by sowing. This technique allows the crop to germinate in almost weed-free environment.

- Crops germinate in weed free environment
- Highly effective in controlling weeds in crops planted in rain less period with the help of irrigation.

Cover Crop

Cover Crop means plants or a green manure crop grown for seasonal soil protection or soil improvement. Cover crops help control soil movement and protect the soil surface between crops. Cover crop reduces wind erosion by shielding the soil with vegetation and anchoring the soil with roots. In India, green manure crops like Sunnhemp, Cowpea, Kolingi, Clover, Lupins etc. are more commonly used. Legume cover cropping in grape, mango, guava and other fruit crops is becoming a common practice in the management of orchards. Cowpea and French beans grow well under guava and sapota tree. In some places to prevent soil erosion, certain permanent cover crops like *Calapogoniu mmuconoides, Centrosema pubescens* and *Peuraria phaseoloides* are raised in the alley spaces especially in Coconut gardens. They are leguminous crops, establish in a short period, dry up during summer to conserve moisture. With summer showers they come up again because of their profuse seeding habit and spread themselves as a vegetative mat by the time the heavy monsoon starts pouring in. Such permanent cover cropping is a common feature in rubber plantations of Kerala and Kanyakumari district of Tamil Nadu.

Intercropping

Intercropping involves growing a smother crop between rows of the main crop. Intercrops are able to suppress weeds. However, the use of intercropping as a strategy for seed control should be approached carefully. The intercrops can greatly reduce the yields of the main crop if competition for water or nutrients occurs.

Crop rotation

Crop rotation involves alternating different crops in a systematic sequence on the same land. It is an important strategy for developing a sound long term weed control program. Weeds tend to thrive with crops of similar growth requirements as their own and cultural practices designed to contribute to the crop may also benefit the growth and development of weeds. Monoculture, that is growing the same crop in the same field year after year, results in a build-up of weed species that are adapted to the growing conditions of the crop. When diverse crops are used in a rotation, weed germination and growth cycles are disrupted by variations in cultural practices associated with each crop (tillage, planting dates, crop competition, etc).

Within a rotation, crop choice will determine both the current and the potential future weed problems that a grower will face. Traditionally, potato (*Solanum tuberosum* L.) was included in the rotation to reduce weed problems before a less competitive crop was grown. For an organic grower, crop choice is complicated further by the need to consider soil fertility levels within the cropping sequence and to include fertility building periods in the rotation. Variations in crop and weed responses to soil nutrient levels can also play an important part in weed management. The inclusion of a fallow period in the rotation in known to reduce perennial weeds. It is best to alternate legumes with grasses, spring planted crops with fall planted crops, row crops with close planted crops and heavy feeders with light feeders.

Choice of crops and varieties

Tall crops and varieties with broader leaves will compete better with late occurring weeds than small varieties with narrow leaves. Some varieties will inhibit and suppress weeds while others will tolerate them. For example, there are witchweed (*Striga* spp.)resistant maize and cowpea cultivars in many countries of Africa, which give better performance at the same level of weeds where other varieties are more affected.

Water management techniques

- Depending on the method of irrigation, weed infestation may be increased or decreased.
- Frequent irrigation or rain during initial stage of crop growth induces several flushes of weeds.
- In lowland rice, where standing water is present most of the time, germination of weeds is less.
- Continuous submergence with 5 cm water results in reducing weed population whereas under upland situation, weed population and weed dry matter is very high.

Pre-germination of weeds

In pre-germination irrigation or rainfall germinates weed seeds just before the cash crop is planted. The newly germinated weeds can be killed by light cultivation or flaming. Pre-germination should occur as close a possible to the date of planting to ensure that changes in weather conditions do not have an opportunity to change the spectrum of weeds (cool vs. warm season) in the field.

Planting to moisture

Another technique similar to pre-germination is planting to moisture. After weeds are killed by cultivation, the top 2 to 3 inches of soil are allowed to dry and form a dust mulch. At planting, the dust mulch is pushed away and large-seeded vegetables such as corn or beans can be planted into the zone of soil moisture. These seeds can germinate, grow, and provide partial shading of the soil surface without supplemental irrigations that would otherwise provide for an early flush of weeds.

Buried drip irrigation

Drip tape buried below the surface of the planting bed can provide moisture to the crop and minimize the amount of moisture that is available to weeds closer to the surface. If properly managed, this technique can provide significant weed control during dry period.

4. Eradicative method (Physical method)

It is the complete elimination of all the live plant parts including seeds from an area. Deep digging and hand weeding are the most useful methods to remove weeds completely. However, these have to be adopted in a limited area.

Deep digging

Digging is labour intensive and slow process to be adopted in a situation where weeds are posing problem because of heavy infestation of root and vegetative propagules deep in the soil. It is having a limited scope but very effective. Deep digging is very useful in the case of perennial weeds to remove the underground propagation parts of weeds from deeper layers of the soil. Digging is accompanied by hand pulling of plant or picking up of roots and vegetative propagules in the soil. This practice allows weed free situation in the first year and impact of weeds in subsequent years is less. Roots of the crop establish gradually gain competitive advantage and thus weed growth is restricted.

Hand weeding

Hand weeding is a slow but very useful method. Here as weeds are pulled up than weeds loose anchorage from the soil, fall dry desiccate and subsequently die. Thus weeds do not compete with the crop. This cannot be recommended for a larger area but can be adopted successfully as a last resort to remove weeds, which have become big and problematic in specific situations.

Hand hoeing

- The entire surface soil is dug to a shallow depth with the help of hand hoes, weeds are uprooted and removed.
- After hand hoeing, the field is subjected to drying to avoid re-establishing of uprooted weeds.

- This method is adopted in irrigated upland crops like finger millet, pearl millet, onion etc.

Digging

- Weeds are removed by digging up to deeper layers so as to remove underground storage organs.
- It is very useful in the case of perennial weeds and it is done with the help of pick axes or crowbars. *Cynodon dactylon* can be effectively controlled by this method.

Mowing

- Mowing is the cutting of weeds to the ground level.
- Mowing is usually practiced in non-cropped areas, lawns and gardens wherein the grass is cut to a uniform height to improve the aesthetic value.
- The common mowing tools are sickle, scythe and lawn mower.

Cutting

- Weeds are cut above the ground surface leaving stubble. It is most common practice against brush and trees.
- Cutting is done with the help of axes and saws.

Dredging and Chaining

- Dredging and chaining methods are used to control aquatic weeds.
- Removing of weeds along with their roots and rhizomes with the help of mechanical force is called dredging.
- The floating aquatic weeds are removed by chaining.
- A very heavy chain is pulled over the water bodies to collect the weeds.

5. General techniques

- Convert weeds into manures / composts
- Mulching reduces weeds
- Weeds can be used as mulches after drying

- Crop rotation reduces weed load
- Closer spacing and judicious water management reduces weed occurrence.
- Biological control of weeds

6. Mulching

Mulching or covering the soil surface can prevent weed seed germination by blocking light transmission preventing seed germination. Allelopathic chemicals in the mulch also can physically suppress seedling emergence. There are many forms of mulches available.

Living mulch

Living mulch is usually a plant species that grows densely and low to the ground such as clover. Living mulches can be planted before or after a crop is established. It is important to kill ad till in, or manage living mulch so that it does not compete with the actual crop. A living mulch of *Portulaca oleracea* from broadcast before transplanting broccoli suppressed weeds without affecting crop yield. Often, the primary purpose of living mulch is to improve soil structure, aid fertility or reduce pest problems and weed suppression may be merely an added benefit.

Live mulch in maize and coconut

Organic mulches

Such materials as straw, bark, and composted material can provide effective weed control. Producing the material on the farm is recommended since the cost of purchased mulches can be prohibitive, depending on the amount needed to suppress weed emergence.

An effective but labor-intensive system uses newspaper and straw. Two layers of newspaper are placed on the ground, followed by a layer of hay. It is important to make sure the hay does not contain any weeds seeds. Organic mulches have the advantage of being biodegradable. Cut rye grass mulch spread between planted rows of tomatoes and peppers was more economic than cultivation.

Fresh bark of conifers and oak as well as rapeseed straw gave good control of weeds when they were laid as mulches under the trees in apples orchards. Materials such as black polyethylene have been used for weed control in a range of crops in organic production systems. Plastic mulches have been developed that filter out photosynthetically active radiation, but let through infrared light to warm the soil. These infrared transmitting mulches have been shown to be effective at controlling weeds.

Straw mulch

Pebble mulch

Here small pebbles like stones are placed on the soil surface. This mulching will be successful in dryland fruit tree culture. The pebbles placed on the basins of trees not only reduce evaporation but also facilitate infiltration of rain water into the basin.

7. Mechanical methods

Mechanical removal of weeds is both time consuming and labor-intensive but is the most effective method for managing weeds. The choice of implementation, timing, and frequency will depend on the structure and form of the crop and the type and number of weeds.

Cultivation involves killing emerging weeds or burying freshly shed weed seeds below the depth from which they germinate. It is important to remember that any ecological approach to weed management begins and ends in the soil seed bank. The sol seedbank is the reserve of weed seeds present in the soil. Observing the composition of the seedbank can help a farmer make practical weed management decisions. Burial to 1 cm depth and cutting at the soil surface are the most effective ways to control weed seedlings mechanically.

Mechanical weeders include cultivating tools such as hoes, harrows, tines and brush weeders, cutting tools like mowers and strimmers, and dual-purpose implements like thistle-bars. The choice of implement and the timing and frequency of its use depends on the morphology of the crop and the weeds. Implements such as fixed harrows are more suitable for arable crops, whereas inter-row brush weeders are considered to be more effective for horticultural use. The brush weeder is mainly used for vegetables such as carrots, beetroot, onions, garlic, celery and leeks. The optimum timing for mechanical weed control is influenced by the competitive ability of the crop and the growth stage of the weeds.

Hand hoes, push hoes and hand-weeding are still used when rouging of an individual plant or patch of weed is the most effective way of preventing the weed from spreading. Hand-weeding may also be used after mechanical inter-row weeding to deal with weeds left in the crop row.

Blind, 'over-the top' cultivation controls very small weeds, just germinated or emerged, before and sometimes after planting. The entire surface of the fields is worked very shallow using flex-tine cultivators (e.g. Lely weeder or rotary hoes, Inter-row cultivations with a rotary hoe in pinto beans (*Phaseolus vulgaris* L.) gave adequate weed control without reducing plant stand or injuring the crop.

The hoe-ridger is specifically designed to achieve intra-row control in sugar beet, Thistle-bars are simple blades used to undercut perennial weeds with minimal soil disturbance. The brush weeder, or brush hoe, is used primarily for inter-row weeding of vegetable crop.

Shallow between-row cultivators such as basket-weeders, beet-hoes, or small sharp sweeps are used to cut off and uproot small weeds after the crop is up. These can get very close to the crop when it's small, without moving much soil into the row, and may be the only tools used on delicate crops like leafy greens, As vigorous crops grown, soil can be thrown into the row to bury in – row weeds using rolling cultivates (e.g. Lilliston), spyder wheels (e.g. Bezzerides), large sweeps or hilling disks. Some of these tools can be angled to pull soil away from the row when plants are small and later turned around to throw soil back on the row during subsequent cultivators.

8. Thermal weed control methods

Flamers

Flamers are useful for weed control. Thermal weed control involves the use of flaming equipment to creale direct contact between the flame and the plant. This technique works by rupturing plant cells when the sap rapidly expands in the cells. Sometimes thermal control involves the outright burning down of the weeds. Flaming can be used either before crop emergence to give the crop a competitive advantage or after the crop has emerged. However, flaming at this point in the crop production cycle may damage the crop. Although the initial equipment cost may be high, flaming for weed control may prove cheaper than hand weeding.

Propane – fuelled models of flamers are the most commonly used. Flaming does not burn weeds to ashes; rather the flame rapidly raises the temperature of the weeds to more than 130 °F. The sudden increase in temperature causes the plants cell sap to expand, rupturing the cells walls. For greatest flaming efficiency, weeds must have fewer than two true leaves. Grasses are difficult to impossible to kill by flaming because the growing point is protected underground. After flaming, weeds that have been killed rapidly change from a glossy appearance to a duller appearance. Flame weeders can be used when the soil is too moist for mechanical weeding and there is no soil disturbance to stimulate further weed emergence.

Flaming can be used prior to crop emergence in slow-germinating vegetables such as peppers, carrots, onion, and parsley. Onions have some tolerance to flaming and flame weeding has been successful in both pre and post-crop emergence conditions and after transplanting. Transplanted cabbage has some tolerance to heat, allowing band flaming to be used along the crop row. Damage can occur when the treatment is applied too early, but the crop usually recovers. In a young pear orchard, where treatments were started on a clean soil after cultivation, flaming kept weed growth in check. In an established apple orchard, there was insufficient control of perennial weeds. Best results are obtained under windless conditions, as winds can prevent the heat from reaching the target weeds. The efficiency of flaming is greatly reduced if moisture from dew or rain is present on the plants. Early morning and early evening are the best times to observe the flame patterns and adjust the equipment.

Freezing

Freezing would be advantageous only where there is an obvious fire risk from flaming. Liquid nitrogen and solid carbon dioxide (dry ice) can be used for freezing weeds. Various test systems using electrocution, microwaves and irradiation have also been evaluated for weed control purposes, but high energy inputs, slow work rates and the safety implications for operators have hampered developments. Lasers have been shown to inhibit the growth the *Eichornia crasispes* (water hyacinth) but did not kill the weed completely. Weed control using ultraviolet light has been patented but remains at an experimental stage.

Infrared Weeders

Infrared weeders are a further development of flame weeding in which the burners heat ceramic or metal surfaces to generate the infrared radiation directed at the target weeds. Some weeders use a combination of infrared and direct flaming to kill the weeds. In general, flame weeders are considered to be more effective because they provide higher temperatures, but burner height and plant stage are important too. Infrared weeders cover a more closely defined area than those of the standards flame weeder, but may need time to heat up.

Soil solarization

Solarization in agriculture would include thermal, chemical and biological changes caused by solar radiation when covered by clear plastic films especially when the soil has high moisture content. The possible mechanisms of weed control by solarization

- Killing of germinating seeds
- Killing of seeds stimulated to germinate in the moistened mulched soil
- Indirect microbial killing of seeds weakened by sub-lethal heating and
- Direct killing of weeds due to heat.

In general, solarization increases heat. This increased heat of temperature induces weed seeds to germinate and get killed.

9. Biological methods

It involves the deliberate use of living organism like insects, fish, disease causing organisms or competitive plants to limit weed infestation.

Biological control would appear to be the natural solution for weed control in organic agriculture.

Table 14. Use of biocontrol agents for weed control

Name of the weed	Bioagent
Cyperus rotundus	*Bactra verutana*
Ludwigia parviflora	*Halticacynea* (Steel blue beetle)
Parthenism hysterophorus	*Zygrogramma bicolarata*
Lantana camara	*Crocidosema lantana, Teleonnemia scrupulosa*
Opuntiadilleni	*Dactylopius tomentosus, D. Indicus* (cochineal scale insect)
Eichhornea crassipes	*Neochetina eichhornea, N. bruchi* (hyachinth weevil) *Sameodes albiguttalis* (hyancinth moth)
Salviniamolesta	*Crytobagus singularis* (weevil) *Paulinia acuminate* (grass hopper), *Samea mutiplicalis*
Alternanthera philoxaroides	*Agasides hygrophilla* (flea beetle) *Amynothrips andersoni*
Tribulus terrestris	*Microlarinus piriformis, M. lareynii*

Table 15. Use of fish & competitive plants for weed control

Name of the weeds	Fish
Lemma, Hydrilla, Potamogeton	Grass carp or white amur
Algae	Silver carp, common carp
Competitive plants for weed control	
Parthenium hysterophorus	*Cassia sericea*
Typha sp.	*Brachiaria mutica*

Bioherbicide approach

Using microbial plant pathogens for weed control. Fungi are more useful for this and hence the term mycoherbicide is also used to refer bioherbicide.

Table 16. Commercial mycoherbicides

Trade name	Pathogen	Target weed
Devine	*Phyophthorapalmivora*	*Morrenia odorata* (Strangler vine) in citrus
Collego	*Colletotrichum gleosporoidesf .sp. aeschynomene*	*Aeschynomene virginica* (northern joint vetch) in rice and soyabean
Biopolaris	*Biopolaris sorghicola*	*Sorghum halepense* (Johnson grass)
Biolophos	*Streptomyces hygroscopicus*	General vegetation (non-specific)
LUBAO 11	*Colletotrichumgleosporoides f.sp. Cuscuttae*	*Cuscutta spp.* (Dodder)
ABG 5003	*Cercospora rodmanii*	*Eichhornea crassipes* (water hyancinth)
Biochon	*Chondrostereum purpureum*	*Prunus serotina*

10. Naturally occurring bio herbicides

Using secondary plant products (allele chemicals) and microbial forms as natural pesticides.

Allelopathy in bio-control programmes

Allelopathy is the direct or indirect chemical effect of one plant on the germination, growth or development of neighbouring plants. I is now commonly regarded as component of biological control. Species of both crops and weeds exhibit this ability. Allelopathic crops include barley, rye, annual ryegrass, buckwheat, oats, sorghum, sudan sorghum hybrids, alfalfa, wheat, red clover, and sunflower. Vegetables, such as horseradish, carrot and radish, release particularly powerful allelopathic chemicals from their roots. Suggestions have been made that allelochemicals and other natural products or their derivatives could form the basis of bioherbicides. However, it is unclear whether the application of natural weed killing chemicals would be acceptable to the organic standard authorities.

The allelopathic effect can be used to an advantage when oats are sown with a new planting of alfalfa. Allelopathy from both the alfalfa and the oats will prevent the planting from being choked with weeds in the first year. Buckwheat is also well known for its particularly strong weed suppressive ability. Planting buckwheat on weed problem, fields can be an effective clean-up technique. Some farmers allow the buckwheat to grow for only about six week before ploughing under. This not only suppress and physically destroys, weeds; it also release phosphorus and conditions the soil.

Use of cover crops for bio-control

- Parthenium- incorporated into soil-Reduction in growth of *Cynodon dactylon*
- Dry plants of Menthi (Cumin)-Leachets controls weeds
- Velvet bean- Suppress purple nutsedge

Use of allelopathic chemicls as natural herbicides

- Xanthotoxin-Inhibits germination & growth of *Lactuca sativa* AAL toxin- *Alternaria alternata lycopersici* (Pathogen) - Effective against dicots at low concentration.

What other factors might need to be taken into account?

Before using allelopathy in weed management programmes there are a number of other factors that might be important in any given situation.

Varieties

There can be a great deal of difference in the strength of allelopathic effects between different crop varieties

Specificity

There is a significant degree of specificity in allelopathic effects. Thus, a crop which is strongly allelopathic against one weed may show little or no effect against another

Autotoxicity

Allelopathic chemicals may not only suppress the growth of other plant species, they can also suppress the germination or growth of seeds and plants of the same species. Lucerne is particularly well known for this and has been well researched. The toxic effect of wheat straw on following wheat crops is also well known.

Crop on crop effects

Residues from allelopathic cropscan hinder germination and growth of following crops as well as weeds. A sufficient gap must be left before the following crop is sown. Larger seeded crops are affected less and transplants are not affected

Environmental factors

Several factors impact on the strength of the allelopathic effect. These include pests and disease and especially soil fertility. Low fertility increases the production of allelochemicals. After incorporation the allelopathic effect declines fastest in warm wet conditions and slowest in cold wet conditions

Allelopathic weeds?

Several weed species have been reported to show allelopathic properties. They include couch grass, creeping thistle and chickweed. Where they occur together they may have a synergistic negative effect on crops

11. ITK's in weed management

- To control *Cyanodon dactylon*, harvested dried stalks of cumin crop are spread in the field. As the stalks decompose and mix with the soil, the weed is destroyed (Farmers in Gujarat)
- Farmers change the variety of paddy crop in each season to control the weeds like 'dhakura' (Farmers in Uttaranchal)
- Manure made with mango leaves is applied in the fields to control *Cyperus rotundus* (Farmers in Tamil Nadu)
- Use of multivariate seeds (MVS) for mixed sowing navathaniyam as intercrops.

12. Navathaniyam

- A multivarietal seed mixture of nine crops viz., two cereals, two pulses, two oilseeds, two spices and condiments and a N fixing green manure ploughed *in situ* as green manure (Somasundaram, 2003).

Control of Nut grass weed in the crop field

- Nut grass is locally called as 'Korai'. If the weed occurs then it will affect the growth of the field crop and it also spread to the entire garden.
- In order to prevent the nut grass ploughing the land has to be performed by using neem plough (plough made of neem wood).Besides, the application of neem cake as organic manure is believed to control nut grass weed in the field efficiently.

13. Recent approaches

- Use of rice bran and application of tamarind seed powder in paddy ecosystem
- Weeds composting
- Relay sowing etc.,

14. Herbicide / Kalaikolli

Cow's urine 250 ml + salt 100 g
Mix it and spray
Useful for weed control out of field

Nematode management

Realizing that nematodes cannot be eliminated, and that we must live with them, the overall goal is to keep the population density as low as possible. The economic threshold is the density of nematodes where the losses incurred exceed the cost of nematode management. It is not advisable to depend on a single method to control nematodes. Efficient management requires judicious and careful integration of several which will result in reduction of nematode populations.

The nematode control methods are

1. Regulatory (Legal) control
2. Cultural control
3. Physical control
4. Biological control

1. Regulatory control

Regulatory control of pests and diseases is the legal enforcement of measures to prevent them from spreading or having spread, from multiplying sufficiently to become intolerably troublesome. The principle involved in enacting quarantine is exclusion of nematodes from entering into an area which is not infested, in order to avoid spread of the nematode Quarantine principles are traditionally employed to restrict the movement of infectedplant materials and contaminated soil into a state or country. Many countries maintain elaborate organizations to intercept plant shipments containing nematodes and other pests. Diseased andcontaminated plant material may be treated to kill the nematodes or their entry may be avoided. Quarantine also prevent the movement of infected plant and soil to move out to other nematodes free areas.

Exclusion

Crop losses caused by plant parasitic nematodes can be avoided through preventing the introduction of specific nematodes or nematode problems in areas wherethe species do not exist. The focal point of exclusion is the target nematode species.Exclusion procedures should be used as first order defenses to prevent dissemination and establishment.

Exclusion procedures include sanitation, certified plant material, nematode free soil or planting media, population reduction or eradication procedures and regulatory activities. Quarantines are used to prevent or slow the spread of plant parasitic nematodes. Certified plant material and nematode free planting media or equipment are used for nematode exclusion.

2. Cultural control

Cultural nematode control methods are agronomical practices employed in order to minimize nematode problem in the crops.

Selection of healthy seed material

In plants, propagated by vegetative means we can eliminate nematodes by selecting the vegetative part from healthy plants. The golden nematode of potato, the burrowing, spiral and lesion nematodes of banana can be eliminated by selecting nematode free plant materials. The wheat seed gall nematode and rice white tip nematode can be controlled by using nematode free seeds.

Adjusting the time of planting

Nematode life cycle depends on the climatic factors. Adjusting the time of planting helps to avoid nematode damage. In some cases crops may be planted in winter when soil temperature is low ad at that time the nematodes cannot be active at low temperature. Early potatoes and sugar beets grow in soil during cold season and escapes cyst nematode damage since the nematodes are not that much active, to cause damage to the crop during cold season.

Fallowing

Leaving the field without cultivation, preferably after ploughing helps to expose the nematodes to sunlight and the nematodes die due to starvation without host plant. This method is not economical.

Deep summer ploughing

During the onset of summer, the infested field is ploughed with disc plough and exposed to hot sun, which in turn enhances the soil temperature and kills the nematodes.

For raising small nursery beds for vegetable crops like tomato and brinjal seed beds can be prepared during summer, covered with polythene sheets which enhances soil temperature by 5 to 10ÚC which kills the nematodes in the seed bed. This method is very effective and nematode free seedling can be raised by soil solarization using polythene sheets.

Manuring

Raising green manure crops and addition of more amount of farm yard manure, oil cakes of neem and castor, pressmud and poultry manure etc enriches the soil and further encourages the development of predacious nematodes like *Mononchus* spp. and also other nematode antagonistic microbes in the soil which checks the parasitic nematodes in the filed.

Flooding

Flooding can be adopted where there is an enormous availability of water. Under submerged conditions, anaerobic condition develops in the soil which kills the nematodes by asphyxiation. Chemicals lethal to nematodes such as hydrogen sulphide and ammonia are released in flooded condition which kills the nematodes.

Trap cropping

Two crops are grown in the field, out of which one crops is highly susceptible to the nematode. The nematode attacks the susceptible crop. By careful planning, the susceptible crop can be grown first and then removed and burnt. Thus the main crop escapes from the nematode damage. Cowpea is highly susceptible crop can be grown first and then removed and burnt. Cowpea is highly susceptible to root – knot nematode and the crop can be destroyed before the nematodes mature.

Antagonistic crops

- Certain crops like mustard, marigold and neem etc have chemicals or alkaloids as root exudates which repell or suppress the plant parasitic nematodes.

- In marigold (*Tagetes* spp.) plants the á – terthinyl and bithinyl compounds are present throughout the plant from root to shoot tips. This chemical kills the nematodes.
- In mustard allyl isothiocyanate and in pangola grass pyrocatechol are present which kills the nematodes.
- Such enemy plants can be grown along with main crop or included in crop rotation.

Removal and destruction of infected plants

Early detection of infested plants and removal helps to reduce nematode spread. After harvest the stubbles of infested plants are to be removed. In tobacco, the root system is left in the field after harvest. This will serve as a inoculum or the next season crops. Similarly in *D. angstus* the nematode remains in the left out stubbles in the field after harvest of rice grains. Such stubbles are to be removed and destroyed and land needs to be ploughed to expose the soil.

Use of resistant varieties

Nematode resistant varieties have been reported from time to time in different crops. Use of resistant varieties is a very effective method to avoid nematode damage. Nemared, Nematex, Hisar Lalit and Atkinson are tomato varieties resistant to *M. incognita*. The potato variety Kufri swarna is resistant to *G. rostochiensis*.

3. Physical control

It is very easy to kill the nematodes in laboratory by exposing the nematodes to heat, irradiation and osmotic pressure etc., but it is extremely difficult to adopt these methods in field conditions. These physical treatments may be hazardous to plant or the men working with the treatments and the radiation treatments may have residual effects

Heat treatment of soil

Sterilization of soil by allowing steam is a practice in soil used in greenhouse, seed beds and also for small area cultivation. Insects, weed seeds, nematodes, bacteria and fungi are killed by steam sterilization. In such cases steam is introduced into the lower level of soil by means of perforated iron pipes buried in the soil. The soil surface needs to be covered during steaming operation. Plastic sheets are used for covering. In the laboratory and for pot culture experimentsautoclaves are used to sterilize the soil.

Hot water treatment of planting material

Hot water treatment is commonly used for controlling nematodes. Prior to planting the seed materials such as banana corms, onion bulbs, tubers seeds and roots of seedlings can be dipped in hot water at 50–55 °C for 10 minutes and then planted.

Washing process

Plant parasitic nematodes are often spread by soil adhering to potato tubers, bulbs and other planting materials. Careful washing of such planting material helps to avoid the nematodes in spreading in new planting field. Washing apparatus for cleaning potato and sugarbeet tubers are commercially developed and are being used in many countries.

Seed cleaning

Modern mechanical seed cleaning methods have been developed remove the seed galls from normal healthy wheat seeds.

4. Biological control

Biological control aims to manipulate the parasites, predators and pathogens of nematodes in the rhizosphere in order to control the plant parasitic nematodes. Addition of organic amendments such as farm yard manure, oil cakes, green manure and pressmud etc encourages the multiplication of nematode antagonistic microbes which in turn checks the plant parasitic nematodes.

The addition of organic amendments acts in several ways against the plant parasitic nematodes. Organic acid such as formic, acetic propionic and butyric acids are released in soil during microbial decomposition of organic amendments. Ammonia and hydrogen sulphide gases are also released in soil during decomposition. These organic acids and gases are toxic to nematodes.

Nematode antagonistic microbes multiply rapidly due to addition of organic matter.Organic amendments improve soil conditions and help the plants to grow healthy. The organic matter also provides nutrition for the crop plants.

For the control of Root knot nematodes: *Meloidogyne incognita & M. enterolobii*

1 kg of bioagent *Purpureocillium lilacinum* may be mixed in 100kg of FYM, mixed well, moistened and stored in shade for 2-3 weeks and applied @ 500 g-1 kg/plant every 3 months

Predacious fungi

Most of the predacious fungi come under the order Moniliales and Phycomycetes. There are two types of predacious activities among these fungi. They are nematode trapping fungi and endozoic fungi.

Non-constricting rings

The trap is formed similar to the constricting ring. It is a non – adhesive trap. The ring becomes an infective structure and kills the nematode eg. *Doctylaria candida*.

In addition to formation of traps and adhesive secretions, the predacious fungi may also produce toxin which kills the nematodes.

Endozoic fungi

The endozoic fungi usually enter the nematode by a germ tuber that penetrates the cuticle from a sticky spore. The fungal hyphae ramify the nematode body, absorb the contents and multiply. The hyphae then emerge from dead nematode. *Catenaria vermicola* often attacks sugarcane nematodes.

Pasteuria penetrans was found to be very effective against the root – knot nematodes in many crops. The *P. penetrans* infested J2 of root knot nematodes ca be seen attached with spores throughout the cuticle.

Rodent and bird pests management

Rat control

- Rats do not live in fields where sheep penning is being practiced.
- Planting closely notchi (*Vitex negundo*) and erukku (*Calotropis gigantea*) around the fields as a fence helps to control rat problem.
- Putting the branches of Thangarali (*Tecoma sands*) around the fields to control rats.
- ‘To control rats in paddy fields, Channampoo (*Cycas circinalis*) flowers are cut in to pieces and placed in many places whose bad odour drives away the rats.
- Pieces of Palmyra (*Borassus flabellifer*) leaves are tied on the poles fixed on the field. The sound produced by the leaves scares away the rats
- Providing owl stands near the rat holes will help in reducing the rat damage.
- To reduce rat population, rat holes are dug and rats are killed after each harvest.
- To catch the rats, a trap made up of wire loops on bamboo pegs is being used.
- Big round shaped earthen pots are buried on the field at ground level. Half of the pot is filled with mud slurry on which baiting material is put on a coconut shell. Attracted rats fall inside the pot and they cannot climb up and get killed.
- Use of soaked rice as bait attracts more rats.
- Putting fresh cow dung on both the fields and bunds to reduce rat problem.
- Papaya- 3 an ripened fruits/ acre

Eli viratti (Rat control)

- Palymyra frond is tied on the stick and kept in the rice field. It will create sound due to air movement, which will scare the rat away.
- Total numbers required: 15 acre

Eli kattuppaduthuthal (Rat control)

- Cow dung balls are immersed in kerosene and kept in 8-10 places of the field randomly. Odours comes from kerosene and cow dung will control rat damage.
- By leaving space (nearly 1 ft) in all four sides of field, rat damage will be reduced.
- Pots are filled with water and buried in the bunds of the field. When rat enter into the field, it will be trapped and killed.

Neekalpoduthal (Rat control in rice field)

- Neekal poduthal is nothing but giving one feet space in 4-5 places of the field randomly when the crop is at milky stage in the rice.

Rat control method in coconut

Developed method of rat control includes an old bamboo basket, binding wire, plastic thread, snap trap and desiccated coconut pieces. An old bamboo basket is tied at four corners with binding wire. All these four binding wires are connected to a single plastic thread and put on a coconut frond which can be pulled up or down. Desiccated coconut piece is attached to a snap trap and placed inside the bamboo basket. Rats get attracted to desiccated coconut and they get locked inside the trap. Dead rats are removed and buried in the soil so as to prevent the spread of diseases.

Extract of *Ipomoea fistulosa*

Leaves of *Ipomoea* are boiled in water and filtered. Sorghum grains are boiled in this extract and placed near the rat burrows. It is believed that rats die after eating it.

Rat control in field

Soak the dead rats (4-5 nos) in field in cow's urine for a week time and sprinkle the decayed rat syrup in all the borders of the field.

Birds perching sticks

In the paddy field few places long sticks are planted and straw rolls are placed to over them, so as to attract to perch over on them. During night hours some predatory birds like owls will come and perch over the sticks. The field rats that are found in the vicinity are caught as prey and eaten.

Prevention of Rodents

- To prevent the damage by field's rats and squirrels in coconut trees, tar is heated and applied to the base of the palm all around upto a height of 0.5 metre from the ground level to a thickness of 15 cm.
- As the tar applied surface will be smooth, the field rats and squirrels cannot climb over it.

Control of Field Rats

- All around the paddy field, "Thalai" (*Pandanus fascicularis*) leaf sheath are fixed closely in the field hedges as fence to prevent entry of rats. The thorns found in the leaf margin will pierce into the abdomen of the rats.
- Keep mud pot filled with straw bits with a hole in the bottom. The pot is placed over the live burrow and by burning the straw pits, smoke will be formed which is directed into the burrow to kill rats. The pot is tightly closed with its lid.
- A mud pot is placed with its bottom hole. Either bamboo or banana bank or "Korai' (nut grass) stem is inserted up to the bottom and by frequent pulling up and down a typical sound is made, which will frighten the rats to leave off the place.

Using Mud pots

A small mud pot with a lid is used as a trap to catch field rats. Mud pot is erected over tripped of wooden sticks at a height of 9 inches from ground and set at certain angle (slanting position). The lid of the pot is put in the field ground, top side facing down and fixed with 3 small sticks.

Inside the pot small quantity of dried fish / fried kernels of ground nut kept as bait. When the rat attempts to get the bait by jumping the pot topple down over the lid and the rat gets trapped inside and subsequently removed and killed. Daily up to 8 rats trapped by this method.

Bird scaring

- Stones are thrown with the help of 'Kavan' made by tying a long piece of gunny thread to both ends of a small piece of leather.
- Stones are thrown with the help of a leather rope called 'Kavattai' to scare the birds.
- Tying the carcass of a crow to a long pole and placing it in the center of field.
- Tying a black cloth to a long pole and placing it in the center of fields to drive away the crows.
- Waste magnetic tapes taken from an audio cassette are tied in the fields in a criss cross manner which the birds will think it as a net.

Paravaithangi (Bird perch)

- Coconut branches are kept in the field randomly
- 'T" shaped stick is used
- Total numbers required : 15 numbers/acre

Tying palmyra fronds

- Dried palmyra fronds are tied to the poles which makes sound when wind blows.
- Total numbers required: 5-10 numbers/acre to keep sparsely.
- Thus, by producing sound, birds are scared.

Tying unused recordable tapes

- Unused recordable tapes are tied across the field in a criss-cross manner and it looks like ant spread over the field.
- Birds perceive that net is bring spread to trap them and avoids to land on that field

Tying polythene sheet and beating drums

- Polythene sheet/papers are tied to a pole as like that of dried palmyra fronds and kept in the field. When wind blows, sound is produced and this will scare the birds.

- To scare the birds, sound is produced by beating drums or by beating a plate with stick.

Use of small flags

- The use of ''small flags'' made of plastic or paper sheets and strings attached to a long rope or plastic twine are examples of materials commonly used by local farmers to prevent birds from feeding on mature rice grains.
- These small flags are placed across the rice fields before harvest.

Use of small colored flags

Inverted coconut fronds

- Coconut fronds are placed in strategic locations in the rice field in an upside down position resembling an owl or a cobra to prevent rodent infestation.

Inverted coconut fronds in rice field

Owl perches

- 'T' shaped wooden sticks (3 feet height) are to be placed in the center /corner of the field at half kilometer distance.
- These sticks are surrounded by paddy straw on top portion act as a stand / resting places for owls in the night hours.
- During the night time, owls would rest on this stick and catch rodent.
- Place 20 "T" shaped wooden sticks per acre.

Bird scaring in maize using tin and stick

- The labour or the farmer engaged in beating the tinbox moved around the field for effective scaring of birds to distance

Wild boar control

Broadcast human hair waste (1-2 cm length) on the pathway of the wild boar route and all possible places of entry. In having the hair bits sticking in the wet nostril (nose) wild boar will rub their face on hand and will ran away.

Chapter 8

ITK's in Organic Farming

Agricultural scientists and policy makers have understood that continuation of modern agriculture might lead to severe ecological and economic problems. We are also convinced that modern agriculture may not be able to meet the requirements of the ever increasing population in the future. So, we are searching for alternative technologies. Several alternatives have been proposed such as low external input agriculture, sustainable agriculture, organic farming, biodynamic farming etc. However, they require, some times little or considerable external inputs whose availability may be uncertain in future.

Hence for the developing countries, the other alternative viz., traditional methods have special advantages over modern agricultural techniques. Also the capital and technological skill requirements in the use of traditional technologies are generally low and their adoption often requires little restructure of the traditional societies. The traditional technologies are nothing but indigenous technical knowledge. By adopting such indigenous knowledge, our ancestors did not face any problem of large-scale pest out break or economic crisis unlike the today's farmers.

Indigenous Technical Knowledge (ITK)

Is the systematic body of knowledge acquired by local people through the accumulation of experiences, informal experiments, and intimate understanding of the environment in a given culture. Learning from ITK can improve understanding of local condition and provide a productive context for activities designed to help the communities.

In addition, the use of ITK's assures that the end user of specific agricultural development projects are involved in developing technologies appropriate to their needs'. Yet, ITK is still an underutilized resource in the development activities. It needs to be intensively and extensively studied, and incorporated into formal research and extension practices to make agriculture and rural development strategies more sustainable.

Though the indigenous technical knowledge (ITK) is region specific, it can be applicable to similar agro-climatic conditions because most of the indigenous agricultural technologies have got scientific rationale. Now the need has come to re-examine and then gradually re-introduce the effective traditional technologies of crop production. However, one may doubt about the productivity of the indigenous practices and varieties. These fears could be dispelled by the reports that the paddy productivity in the Chengalpattu area of Tamil Nadu during 1780's was more than 2500 lbs / acre. Comparatively in 1960's, the productivity was only 1680 lbs / acre. Similar data exist for North Arcot, South Arcot, Tanjore and Coimbatore districts of Tamil Nadu, for the late 18th century to early 19th century.

However, many of our indigenous practices in agriculture and allied fields have been replaced by the so-called modern technologies and they have become obsolete, especially among the younger generations. Now these indigenous practices are endangered ones and these is a possibility for them to become extinct particularly during this era of globalization, liberalization and commercialization.

Therefore, there is a need to systematically document the ITK as they are the unwritten body of knowledge. There is no systematic record to describe what is, what it does, how it does, means of changing it, its operations, it boundaries and its applications. It is held in different brains, languages and skills in as many groups, cultures and environments. Tamil Nadu is also a treasure land of indigenous practices in agriculture. However, only a few attempts have been undertaken by social scientists to document the available indigenous agricultural practices but none has evaluated scientifically.

Rational and Principles of ITK's

The ecologically sound designs or elements of indigenous practices, which have been losing importance in high technology production system, must be saved and synthesized appropriately to attain sustainable farming. The total eco-design involves, combing local elements with innovation based traditional methods with newer elements to form a balanced whole. In order to encourage and consistently apply this indigenous knowledge, which are degenerating, needs scientific back up. Though the ITK's are region specific, once the advantages are ascertained, the principle underlying the practices may be extrapolated to similar agro-climatic conditions, by carrying out the experiments using the materials locally available in the farm holdings. An agricultural technology based on Indigenous knowledge may bring moderate to high levels of productivity using local resources. Given favourable political, social and ecological conditions, such agro-technologies may be sustainable at a low cost for a long period. Also, in our National agricultural policy, concerted efforts were made to pool, distill and evaluate traditional practices, knowledge and wisdom on organic farming and to harness them for sustainable agricultural growth. Efforts are being made as progrmames for utilization of domestic and agricultural waste for organic matter repletion and pollution control.

All ITK's go by the principle of 'permanence'. It is not so with modern technologies with synthetic inputs. All the ills in crop production (soil mining, degradation, pollution, etc.,) are due to decreasing attention to ITK. Indigenous practices in agriculture are organic in nature. They do not cause any damage to the air, water and soil, safe to the human beings and are free from causing environmental pollution. Therefore, now the need has come to re-examine and then re-introduce the effective traditional methods of crop production and protection using organic sources, because there is considerable demand and scope for development of organic technologies either individually or as a package, without necessarily aiming at full adoption of organic system.

Indigenous agricultural practices can play a key role in the design of sustainable and eco-friendly agricultural system, increasing the likelihood that the rural populations will accept, develop and maintain innovations and interventions. In *Declaration statement* of the 88th session of Indian Science Congress held in New Delhi, 2001 it is opined that if the modern techniques are integrated with the traditional and indigenous practices, that will alleviate the poverty and results in the prosperity of the country. Many indigenous practices documented illustrate how well the farmers in the tropics learned to manipulate and derive technologies from local resources and natural processes, applying the principles of agroecology without knowing that this term exists.

Soil and water management

- For soil improvement in 'theri' lands of Tuticorin district, 200 tones of tank silt are applied per acre followed by 50 tonnes per year for the next few years.
- About 10 kg. of neem cake is soaked in 10 lit. of cow urine along with ½ kg. of waste asafoetida and left over night. In the next day, it is sprayed for 1 ac. after dilution as liquid manure.
- Outer shells of tamarind fruits are applied in the field to control *Cyperus rotundus*.
- A mixture is made with 1 lit. of neem oil, 3kg. of fine sand and 3 kg of cow dung and heaped in shade covering with a moist sack for 3 days. On the fourth day, the mixture is dissolved in 150 lit. of water and sprayed to control all sucking pests.
- About 10kg. of dried cow dung is ground into fine powder and mixed with ash (obtained from brick kiln) dusted in the early morning, to control pests and diseases.
- Garlic acts on a wide spectrum of organism in unrelated crop plants singly or in combination with neem products, chilli, asafetida etc. Besides garlic is effective against bacteria, fungi and nematodes.
- *Calotropis* leaf extract is applied at the place of termite attack to control them.
- Consolidation of lands gives better results even they are less fertile.

- Fields which are nearer to rivers will give lesser yield
- Better yield is obtained from the wetlands near water sluice of canals and the dry lands near foothills.
- Laying stone bunds around the fields across the slope for preventing soil erosion and for conserving moisture.
- Planting vettiver (*Vetiveria zizanoides*) slips across the slope or around the fields to prevent soil erosion.
- To minimize soil erosion perennial vegetation is grown on the field bunds.
- New garden land and old wetland will yield better.
- Intensive care is required for wetland crops as compared to the garden land crops.
- Waterlogged dry lands are unsuitable for cultivation.
- Soil character decides the choice of crops for cultivation.
- Red soil is suitable for continuous cropping.
- Black soil has more water holding capacity than the red soil.
- Sandy soil is less suitable for the cultivation of many crops.
- Excessive application of farm yard manure (FYM) improves the soil texture.
- Tank silt is applied to increase the soil texture.
- Manures and fertilizers are applied based on soil character.
- If the weed growth is profuse after the rains, it indicates high soil fertility.
- However the low soil fertility is indicated by the growth of the weed Aduthinnapalai (*Aristolochia bracteolata*).
- Addition of red soil to black soil increases the fertility of the black soil and vice versa.
- Practicing sheep/cattle penning during summer season to improve the soil fertility.
- Practicing mixed cropping or inter cropping of legumes in rain fed areas to maintain the soil fertility.

- Cultivating Kolinji (*Tephrosia purpurea*) in between the fruit trees in sloppy lands to prevent soil erosion and to improve soil fertility.
- High moisture content in the soil is identified with the occurrence of 'Nuna' tree (*Morinda tinctoria*).
- For moisture conservation deep Ploughing is done during summer.
- Land is well ploughed and powdered to conserve more moisture.
- Application of tank silt (taken from black soil tanks) on the red soil fields to increase the water holding capacity of the red soil.
- Raising and ploughing daincha (*Sesbania* spp.) and sun hemp (*Crotalaria juncea*) inthe field before flowering increase water holding capacity at the soil.
- It is better to grow sorghum, finger millet and chilies if the water is saltish.
- Irrigation is given to any crop at a stage that while walking on the fields, our foot should not create any print on the soil, which is taken as the indication.
- Growing 'Poovarasu' (*Thespesia populnea*) tree near the wells reduces water loss through evaporation.
- Wetlands having 'Aarai' weeds (*Marshilea quadrifolia*) and garden lands having 'Arugu' (*Cynodan dactylon*) weeds give better yields.
- Red soils having 'Arugu(*Cyanodandactylon*) weeds and black soils having nut grass (*Cyperus rotundus*) weeds are the best of their kind.
- Sowing densely the daincha (*Sesbania* spp.) green manure and ploughing in-situ at its flowering to correct alkaline soils.
- Growing sun hemp (*Crotalaria juncea*) in alkaline soils and ploughing in-situ before flowering to rectify its alkalinity.
- Application of 'Pirandai' (*Cissus quadrangularis*) to reduce alkalinity.

- Neem leaves are applied to correct alkalinity.
- Application of shells of neem seed to reduce salinity in soils.
- Application of neem cake to correct salinity.
- Palmyia (*Borassas flabellifer*) leaves are cut into pieces and applied in large quantity to correct alkalinity.
- To correct alkaline soils, pungam (*Pongamia pinnata*) leaves, or outer shells of tamarind fruits are applied.
- Mixing and applying coir waste with compost to correct alkalinity.
- Application of sugarcane bagasse and sediment after extraction of country sugar to correct alkaline soils.
- Putting leaves and branches of Indian gooseberry (*Phyllanthus distichus*) in the wells reduce salinity in water.

Preparatory cultivation

- Achieving fine tilth is better than applying manures.
- It is better to plough intensively than extensively.
- It is better to have deep ploughing rather than shallow ploughing for good crop growth.
- Plough four times for garden land and seven times for wetland.
- Summer Ploughing gives good crop in the ensuing season.
- Tying paddy straw around the sole of country plough to make a wider furrow while forming broad bed and deep furrows.
- Garden lands are ploughed deep to conserve more moisture.

Manures and manuring

- Commonly used green leaf manures are Kolingi (*Tephrosia purpurea*), Calotropis (*Calotropis gigantea*), Nuna (*Morinda tinctoria*), Pungam (*Pongamia pinnata*), Neem (*Azadirach taindica*), Poovarasu (*Thespesia populnea*) and Adathoda (*Adhathoda vasica*).
- Green manure crops like daincha (*Sesbania* spp.), kolingi (*Tephrosia purpurea*) sunn hemp (*Crotalaria juncea*) etc., are raised and ploughed *in situ* before their flowering.

- Red gram is also used as a green manure crop which improves the soil fertility.
- Applying water hyacinth plants either as a compost or as burnt ash to the fields for supplying potash.
- Sheep penning results in more crop yields.
- *Goat* manure gives benefits to crops grown in the same season.
- Goat manure is good for the first season and cattle manure and green manure for the second season.
- Poultry manure serves as a good source of crop nutrients.
- Cow urine is more nutritious than cow dung.
- Soil fertility can be better increased by the use of pig dung than by the cow, sheep or goat waste.
- It is better to apply cattle manure for garden land and dry land and leaf manure for wet land.
- Near the irrigation channel, a pit is dug in which, cow dung, foliage of *Calotropisgigantea,* neem cake powder and cow urine are applied, mixed well and allowed to decompose. Then it is allowed to mix with the irrigation water, to supply the nutrients, and to control pests and diseases. Tank silt is applied every year in dry lands for better yields. Termite hills serve as good manure.
- Adding sand from an ant hill to the field gives good yield.
- Foliar spray of manures given on full moon day yields better results.

Weed management

- Weeding is not required under dry land condition. If weeding is not done the weed growth is controlled naturally and it also helps to conserve moisture.
- Repeated ploughing will reduce weed population.
- Crop yield will be less in the fields having 'Arugu' (*Cynodon dactylon*) weeds.
- To control "Arugu (*Cynodon dactylon*) grass" in black soils the field is kept fallow for 3 years.

- Cultivating rice once in three years in garden lands to control 'Arugu' (*Cynodon dactylon*) weeds.
- Raising and ploughing the green manure crops tike (*Sesbania spp.*), kolingi (*Tephrosia purpurea*) in the field before their flowering to reduce weed population.
- Raising *Calotropis gigantea* as a green manure to check the growth of Aarai (*Marsilea quadrifolia*) weed.
- Growing horse gram to control nut grass (*Cyperus rotundus*).
- Growing cowpea as a green manure to control nut grass.
- Allowing swine in the fields to eradicate nut grass.
- Frequently ploughing the fields by wooden plough made up of neem trees and frequent application of neem cake in the soil to control nut grass.
- Dissolving 1 kg of salt and 100g sarvodaya soap in 10 litre water and spraying this solution to control all the weeds except nut grass.
- To control nut grass in the field 50 kg neem cake is applied both at the time of ploughing and sowing.
- Dissolving 200 g salt in 1litre water and spraying to eradicate congress weed (*Parthenium hysterophorus*).
- Continuous submergence of field for some time controls the weeds.
- Keeping the irrigation channels free from weeds,

Pest and disease management

- Uninterrupted drizzling of rain leads to the occurrence of more number of pests and diseases.
- Winds blowing at the end of December bring lot of pests.
- Crops grown in alkaline soils are more prone to disease attack.
- Small lamps are placed on either side of the house entrance and light from the lamps acts asa light trap and the farmers are able to identify the pest outbreak.
- To prevent the attack of aphids and mite flies, sorghum or pearl millet is grown very closely in 4 rows around the fields to act as a shelter so that these pests can not enter the fields.

- Growing Thangarali' (*Tecoma stands*) and 'Sevvarali' (*Nerium oleander*) as border crops, which act as trap crops and control the insect liest.
- Thiruneeru' (Sacred ash) is dusted on the crops to reduce pest attack.
- Kitchen ash is applied to control aphids.
- To control the sucking pests, 5 kg tobacco powder is soaked in a mixture of 10 litre cow urine and 5 lit water for 5 days. Then it is filtered and diluted with 80 it water and sprayed.
- Mixing cow urine, neem oil and tobacco decoction together and spraying on crops controls sucking pests.
- Burning of rice stubble and straw effectively reduces stem rot caused by *Sclerotiumoryzae* and sheath blight caused by *Corticium sasakii*. Flaming was effective in controlling *Verticillium dahliae* in peppermint. The only disadvantage is that the beneficial organism may also be destroyed.
- Leaves of *Calotropis gigantea* and *Strychnos nux-vomica* and neem cake are soaked in water in a mud pot and fixed in the field. Moths are attracted towards the smell, fall inside and die.
- Leaves of notchi (*Vitex negundo*) and pungam (*Pongamia pinnata*) *are* also used to control moths.
- About 5 kg *Calotropis* leaves is soaked in a mixture of 10 liter cow urine and 5 litre water for 5 days. Then it is filtered and diluted with 80 li. of water and sprayed to control defoliation.
- Spraying sarvodaya soap solution to control mealy bugs.
- To control nematodes, pungam (*Pongamia pinnata*) or iluppai (*Bassia latifolia*) cakes are applied.
- Cow dung, Cow urine, *Calotropis* leaves and neem cake are put in a pit near the irrigation channel. After decomposing, it is mixed with irrigation water.
- Grinding the leaves of *Calotropis gigantea* with the fruits of *Datura metal,* soaking in water for 15 days, filtering and spraying to control all the pests

- Two handful each of leaves of thumb (*Leucasaspera*), kuppaimeni (*Acalypha indica*), thulasi (*Ocimum canum*). *Datura metel,* neem, nochi (*Vitex negundo*), 5 fruits of *Datura metal* and handful each of neem cake and Iluppai (*Bassia latifolia*) cake are pounded together and soaked in water in earthen pot for 10 days. Then it is filtered, diluted (100 ml./lit.) to which 100 ml sarvodaya Khadi soap solution and 100 ml neem oil added and sprayed to control all insect pests.
- Leaves of Tulsi (*Ocimum canum*), seeds of *Nerium oleander and* fruits of *Datura metal* are taken in equal quantities, powdered and soaked in cow urine for 10 days. Then it is filtered and diluted (100 ml/lit.) to which 100 ml neem oil is added and sprayed to control all insect pests.
- Neem oil and neem seed kernel extract are the general organic pesticides used to control many pests.
- During the night time on full moon day of Tamil month 'Karthigai' (Nov.-Dec.) 'Chokkapanai' (Community firing) is performed as a part of celebrations in a common place in the village by which the pests get attracted and killed. Ash from this fire is dusted on the crops to control sucking pests.
- Crop wastes are burnt and its ash is dusted on the fields to control diseases.
- If neem cakes are applied as basal fertilizer there will not be any incidence of diseases.
- One kg leaves of seemaikaruvel (*Prosopis juliflora*) is pounded and diluted with water and sprayed to control yellow mosaic virus.
- Spraying cow urine to control many pests.
- A mixture of extracts of garlic and neem cake is *sprayed* to control aphids.
- Planting 'Pirandai' (*Cissus quadrangularis*) vines around fields to protect against termites.
- Grow castor on the fields to control termites.

- Spread neem leaves over the nursery to control termite damage.
- Putting neem cake inside a gunny bag and placing it in the irrigation channel controls mites.
- Termites destroy the seedlings in nursery grown in dry land condition. To control these termites, apart from putting the neem leaves, sheep wool and human hairs are also put. Termites eating these hairs die.
- Pouring decoction of finger millet roots on the root zone of crops to control termites.
- Before planting tree seedlings, dried leaves and trashes are burnt in the pits to protect the seedlings against termite attack.
- Dusting ash in the pits before planting tree seedlings also helps to prevent termites.
- Sprinkling 5% common salt solution to reduce termite attack on the trees.
- After the harvest of tobacco leaves, their stems and roots are ploughed in-situ to control the termites
- Tobacco soaked water is poured on the ant mounds to control them.
- Any spraying is to be done in the early morning.
- Take 30gm. of grounded *Nerium* (Arali) seeds in 10 lit. of water for 1 hour and mix with khadi soap and spary to control thrips, aphids, whiteflies and leaf eating caterpillars.
- Clear polyethylene placed over moist soil, during summer days raises the temperature at the top 5 cm of soil to as high as 52°C. The increased soil temperature from solar heat, known as solarization inactivates many soil borne pathogens and reduces the inoculum and the potential for disease

Storage Pest Management

- Lime juice is mixed with grains and then sun dried before storage to prevent insect pests
- Seeds are safely stored in earthen pots after mixing with the leaves of neem and *Vitex negundo*

Rainfall

- Rain water received during the 'Magha' constellation is stored and used as a growth promoter on the standing crop
- Large number of fireflies seen at night on the forest trees is a sign that the monsoon will start early

If there is rain, accompanied with lightning and mild thunder on the second day of Jayastha month (May – June), there will be no rain for the next 72 days.

Chapter 9

Organic Crop Production Techniques

Organic farming is a crop production method respecting the rules of the nature, targeted to produce nutritive, healthy and pollution free food. It maximizes the use of on -farm resources and minimizes the use of off – farm inputs.

The general guidelines on organic production of crops are prepared based on National Programme for Organic Production (NPOP) launched by Government of India. These guidelines enable the growers to attain more or less the same level of the productivity of conventional farming within a few years and at the same time maintain the fertility of the soil and protect the ecological balance.

General guidelines for Organic Crop Production

Choice of crops and varieties

All species and varieties that are cultivated should be adapted to the soil and climatic conditions and be naturally resistant to pest and disease of the region. All seeds and planting materials should be from crops of organic cultivation. When organic planting materials are not available, chemically untreated conventional planting materials shall be used initially. The use of genetically engineered seeds, pollen, transgenic plants and plant materials is not allowed.

Conversion period

The establishment of an organic management system and building of soil fertility requires an interim period, the conversion period. The duration of the conversion period will depend upon

- The past use of the land
- The ecological situation

The plant products produced annually can be certified organic when the national standards requirements have been met during a conversion period of atleast two yearsbefore sowing or in the case of perennial crops other than grassland, atleast 3 years before the first harvest of the products. Conversion period can be extended by the certification program depending on the past use of the land and environmental conditions. The accredited certification program may allow plant products to be sold as "produce of organic agriculture in process of conversion" when these national standards stipulation have been met for atleast 12 months.

Diversity in crop production

Diversity in crop production is achieved by a combination of:

- A versatile crop rotation with legumes
- An appropriate coverage of the soil during the year of production with diverse plants pecies.
- Follow crop rotation for annual crops and intercropping for perennial crops.
- Avoid crops belonging to the same family in the rotation.
- Biofencing with green manure shrubs or neem and other plant protection agents.

Manurial Policy

Sufficient quantities of biodegradable materials of microbial, plant or animal origin should be returned to the soil to increase or atleast maintain its fertility and the biological activity within it. Organic material must be the product of organic farms and the farms must become self sufficient in producing such organic material

Soil fertility should be maintained or enhanced by

- Raising green manure crops, leguminous crops
- Incorporate crops residues

- Use biodegradable materials of microbial, plant or animal origin
- Encourage the use of on - farm organic inputs
- Use of synthetic or chemical fertilizers and growth regulators are not permitted
- Mineral based materials like rock phosphate, gypsum, lime etc in limited quantities and in their natural compositions.
- Prevent the accumulation of heavy metals and other pollutants
- Minimize the nutrient loss by management practices
- Apply manures as per soil test results
- Maintain adequate pH levels
- Manures containing human excreta shall not be used (Products for use in fertilizing and soil conditioning are listed in Appendix II).

Pests, diseases and weed management

Organic farming systems should be carried out in a way, which ensures that losses from pests, diseases and weeds are minimized. Conditions for minimizing the loss due topests, diseases and weeds are

- Balanced manurial programme
- Use of crops and varieties well adapted to the environment
- Fertile soil of high biological activities
- Adopt rotations
- Companion planting
- Green manuring
- Natural enemies of pests and diseases should be protected and encouraged.
- Cultivate trap crops

Pest and disease control

- Prohibit the use of synthetic chemicals
- Use preventive cultural techniques
- Encourage and protect natural enemies

- Use products from local plants and of biological origin prepared at the farm.
- Prohibit the use of genetically engineered organisms and products
- Brand name products must always be evaluated

Weed control

- Slash weeding
- Use mechanical weed control
- Use weeded materials as mulch
- Use clean equipments for organically managed areas
- Use of synthetic herbicides, synthetic growth regulators and synthetic dyes are prohibited

Contamination control

All relevant measures should be taken to minimize contamination from outside and within the farm. Accumulation of heavy metals and other pollutants should be limited. That cultivation has to guard against the possibility of pesticide and weedicide contamination and the carriage of inorganic chemicals used as fertilizers by irrigation and drainage. For protected structure coverings, plastic mulches, fleeces, insect netting and silage wraping, only products based on polyethylene and polypropylene or other polycarbonates are allowed. These shall be removed from the soil after use and shall not be burned on the farm land. The use of polychloride-based product is prohibited.

Soil and water conservation

Soil and water resources should be handled in a sustainable manner. Relevant measures should be taken to prevent erosion, salination of soil, excessive and improper use of water and the pollution of ground and surface water. In sloppy lands adequate precautions should be taken to avoid the entry of run off water and drift from the neighboring farms. Clearing land through the means of burning organic matter shall be restricted to the minimum.The clearing of primary forest is prohibited.

Organic Crop Production Techniques

Rice

- Treatment of paddy seeds in diluted biogas slurry for 12 hours increases resistance of seedlings to pests and diseases.
- During panicle formation in paddy, the flowers of *Cycas circinalis* are placed on sticks in paddy fields @ 4/ac. Its unpleasant odor repels ear head bugs.
- About 30 kg of tamarind seeds are applied for an acre of paddy field 1 day after transplanting to boost up the crop growth and yield.
- Soaking the paddy seeds in diluted cow's urine before sowing, considerably reduces the incidence of leaf spot and rice blast
- Pre-soaking of paddy seeds in milk increases its resistance against 'tungro' virus and 'stunt' virus
- For control of red leaf spot disease in paddy, the seeds are soaked in 'Pudina' leaf extract (*Mentha sativa*) for 24 hours
- 'T' shaped bamboo stands are placed in many places in the paddy fields so that birds can sit on them and feed on the larvae and adults of rice pests.
- Sowing on eighteenth day (Aadipperukku) of Tamil month Aadi (Jul-Aug.) ensures good harvest.
- Dhaincha (*Sesbania* spp.) seeds are sown on paddy main fields when paddy nursery is raised and the grown up dhaincha is ploughed in-situ during field preparation.
- Plough the main field for four to six times for better yield.
- Good harvest can be obtained from the crop transplanted during Aavani i.e. Aug. - Sep.
- The crop transplanted during October-November will give reduced yield.
- The rice crop will establish better if it is transplanted along the wind direction.
- Planting the 'samba (Aug) crop thickly and 'navarai' (Feb.) thinly.

- Practice sheep penning during summer to get more yield.
- Practice sheep penning for the first season and green leaf manure for the second season for better yield.
- Apply 100 kg of pig manure for one acre of rice at 10 days after planting to get higher yield.
- Apply the neem seeds @ 40 kg / ac as basal to get more yield as compared to the equal quantity of neem cake.
- Irrigate the fields, allow the weed seeds to germinate and then plough the fields to incorporate the weeds into the soil before sowing or transplanting of rice crop to control weed growth.
- Cultivation of sunhemp or daincha helps to control the nut grass (*Cyperus rotundus*) weed.
- Application of *Calotropis gigantea* as green leaf manure will prevent thrips attack in the nursery.
- Neem (*Azadirachta indica*) oil cake extract is sprayed to control thrips in rice.
- Dragging the branches of country ber or Aloe sp. on the affected field to control the leaf roller.
- Neem oil is mixed with water @ 30ml / lit and sprayed to control stem borer in rice.
- Dusting chullha ash in the early morning to control stem borer and ear head bug.
- To control the ear head bugs, 10 kg cow dung ash is mixed with 2 kg lime powder and 1 kg powdered tobacco waste and dusted on the rice crop during morning hours.
- Hundred ml. of leaf extract of "Karuvel" (*Acacia nilotica*) and 10 kg cow dung are dissolved in 10 lit. of water and sprayed on the rice crop to control ear head bug.
- Growing or planting calotropis at 12 feet interval on all sides of paddy fields to control the hoppers.
- Applying neem cake before last plough to control root rot and nematode problem.
- A mixture of 5 kg common salt and 15 kg. of sand is applied for 1 acre to control brown spot disease.

- Soaking the paddy seeds in 20% mint leaves solution before sowing will control the brown leaf spot.
- Spraying the leaf extract of *Adatoda vasica* to control rice tungro.
- Palmyra (*Borassus flabellifer*) fronds are tied on to poles and kept on the corners of rice fields so that the noise produced by them scare away the birds like ducks, sparrows etc. and save the grains being damaged.
- When one ear head contains about 100 grains, the yield will be 20-22 quintals/ac.
- One hundred and twenty grains found in a rice ear head indicates the full yield.
- Use large mud pots called 'Kudhir' as high as six feet for storing paddy grains for longer periods.
- Putting the leaves of notchi (*Vitex negundo*) and pungam (*Pongamia pinnata*) inside the *Kulumai* to ward off storage pests.
- Mixing the paddy grains with the leaves of pungam (*Pongamia pinnata*) or notchi (*Vitex negundo*) or neem (*Azadirachta indica*) before storage to avoid storage pest attack.

Pulses

Redgram

- Spray the decoction of tobacco waste to control sucking pests and caterpillars.
- Red gram seeds are mixed with red earth slurry, dried and stored to avoid storage pests.
- Castor seeds are fried, powdered and mixed with red gram seeds to reduce pest attack during storage.
- Storing the red gram seeds after mixing them with 'sweet flag' (*Acorus calamus*) powder @1 kg per 50 kg seeds to preserve them for one year.
- Dry the red gram seeds well and store them in gunny bags after placing dried leaves of 'Naithulasi' (*Ocimum canum*)

inside them to prevent pod borer attack. (Also for black gram)

- Putting the pods of dried chillies in the red gram container to control bruchids (beetle) attack.

Blackgram

- When a wooden plank is moved with pressing over the drying gram, splitting of gram indicates optimal drying.
- Bullocks pulling a heavy stone roller are allowed to trample over the harvested black gram crops spread out in the threshing yard so as to separate the grains.
- Yield will be higher in black gram crop, if it is sown in the second fortnight of September.
- Neem oil is sprayed @ 6 lit /ac. to control powdery mildew in black gram crop.
- Mixing the black gram seeds with ash and storing them in earthen pots for longer period (Also for cowpea and green gram).
- Coating the black gram with castor oil to increase the keeping quality.
- Mixing the black gram with sweet flag (*Acorus calamus*) powder for seed purpose.
- Black gram grains broken into halves will escape from weevil attack during storage.

Cowpea

- Putrefied buttermilk is sprayed on cowpea crop to control yellow mosaic disease (Also for green gram).
- Vegetable oil is mixed with cowpea before storage.
- For safe storage, cowpea seeds are filled in earthen pot to its 4/5th volume and the remaining volume is filled with ash (also for field bean)
- Mix cowpea seeds with red earth slurry, dry and store them in earthen pots for one year.

Banana

- Unripened banana bunches are piled in a vessel and incense sticks are inside the vessel. Then if the lid of the vessel is closed, the bunches will ripe in about 12 hours.
- For quick ripening of banana fruits, lime solution is sprinkled over the bunches.
- For easy ripening of banana, neem leaves are inserted in between the bunches
- About 25g mixture of neem oil cake and castor oil cake is applied around each banana sucker 60 days after planting to control.
- Diluted tobacco leaf extract is sprayed on banana crop to control leaf spot diseases.
- Suckers, which are half foot in height and 2½ kg. in weight are used for banana planting.
- Banana suckers are immersed for a while in 1 lit. of neem oil dissolved in 100 lit of water before planting in order to prevent rhizome rot.
- Groundnut cake is applied to banana crop for better yield.
- To control fruit rot in banana during storage, the fruit stalks are soaked in 10% thulasi (*Ocimum canum*) leaf extract or 1 % neem oil solution and stored.
- Banana crop raised after the marigold cultivation invites less nematode attack.
- Growing *Sesbania spp.* (trees) as border crop around banana fields to act as a shelter crop in order to prevent the wind damage.
- Neem leaves are put inside a vessel containing banana hands for ripening of fruits. But ripening will take about four days.

Mango

- Neem oil is sprayed to control the hoppers.
- Sunflower is cultivated in between the mango trees to attract honey bees which increase pollination and fruit production.

- To induce early ripening of mango fruits, they are spread on a layer of the branches of 'Aavaram' (*Cassia auriculata*) plant on the floor and again covered with its branches and finally

Vegetables

Brinjal

- In order to prevent fruit rotting in brinjal plants, a solution is made of 1 lit. of water and eight crushed leaves of *Aloe vera* and sprayed on the crops *Chrysanthemum.*
- *Coronaries* are grown as a border crop in brinjal to control fruit borers.
- Poultry manure is applied for more yields in brinjal
- Grinding and applying the neem seeds @ 40kg./ac. on 35th day after transplanting gives higher yield.
- Growing castor in Brinjal fields as border crop to act as a trap crop for insects.
- Growing onion as intercrop in Brinjal to control many pests including fruit borers.
- Mixing and grinding well neem cake with *Aloe vera* and soaking in water for 10 days, after which spraying the filtrate to control thrips.
- Ash and turmeric powder are mixed in equal proportion and sprinkled to control aphids.
- Sprinkling of lime powder to control mealy bugs.
- Cow urine, neem oil and tobacco decoction are mixed and sprayed to control all sucking pests.
- Spraying neem cake extracts to control mites and the spotted beetle (*Epilachna octopunctata*) in brinjal.

Cucurbits

Snake Gourd

- Soaking the seeds of snake gourd in cow dung solution for 6hour before sowing helps for early germination and withstanding drought conditions.
- Asafoetida is dissolved in water @ 25g per lit and sprayed to control flower dropping.
- After the harvest of snake guard, cabbage foliage is ploughed in-situ to serve as a manure

Ribbed Gourd

- Ribbed gourd having even number of raised ridges on its skin is likely to be sweeter and the one with odd number of ridges is likely to be bitter.

Bitter Gourd

- Bitter gourd germinates faster and grows well when its seeds are soaked in milk for a day prior to sowing

Pumpkin

- Storing of pumpkin seeds in dried fruit storage containers. After completely drying the fresh fruits, the inner pump and seeds are removed. Then the seeds selected for storage are mixed with ash and placed inside the fruit container. The entrance is covered with moist red earth and a small hole is made for aeration.

Bottle Gourd

- Store of seeds by sun drying them with fruit itself, as the dried fruits act as a storage container.
- Soak the seeds in water and wrapping in a moist cotton cloth to enhance the germination of Bottle gourd.
- Practice of stripping the lateral branches induces formation of more number of such lateral branches and ensures good quality gourds.

Cabbage

- Frequent weeding in the cabbage nursery for better establishment of cabbage plants in nursery and main field.
- Frequent weeding in the cabbage main field for getting matured heads earlier.
- Spray of the spray solution made of the mixture of ash and cattle urine on the cabbage plants to control leaf hoppers.
- Application of 1.5 kg lime mixed in 2-3 litres of buttermilk and weekly intervals by using broomstick to control cabbage caterpillar.
- Hoist five-balled sticks in the cabbage fields at an interval of 20' to control insect pest attack on cabbage.

Tuber crops

Radish

- Raise the Radish crop after potato needs little expense because of the loose soil after digging the potato tubers.
- Raise radish as a rainfed crop comes up well by utilizing the available residual moisture in the field.
- Raise radish after the potato requires no manures or fertilizers because of the residual fertilizers and manures available in the field.
- Wash of the harvested radish tubers with water for better shipping.

Beetroot

- Raise Beet root as an intercrop with carrot requires little individual care, as the operations done for carrot itself is enough for beetroot.
- Selection of big seeds having better striving for seed purpose.
- The stripping of beet root tubers indicates good quality.

Carrot

- Crop rotation of carrot with potato, peas to get more income because same next crop will not give more yield.
- Fencing with the thin branches of Seegai tree from the local forest to arrest soil erosion.
- Construction of stone wall across the slope in the field to arrest soil erosion.
- Forking and clod breaking in the main field before taking up cultivation to make the bottom soils come up and top soils go down.
- Spray the solution made from the mixture of ash and cattle urine on the carrot plant to kill leafhoppers.
- Assess the harvesting time by digging sample carrot plants. If the carrot is of orange colour and big size that indicates the maturity.
- Wash the carrot tubers after harvesting in the running water of small streams for better shining of carrot tubers.
- Use the stem and leaves after harvesting either as cattle feed or for making compost to use as manure in the next crop.

Potato

- Crop rotation of potato crop followed by vegetables for the nematode management.
- Grow potato increases soil texture.
- Mixed cropping of potato with Marigold (*Tagetes spp.*) reduces the risk of root nematode.
- Rotation of potato crop with other crop minimizes pest and disease infestation.
- Crop rotation of potato crop followed by cabbage or radish or peas gives good returns to the farmers.
- Crop rotation of potato with carrot or beet root or turnip utilizes the time and moisture available.
- Crop rotation of potato crop followed by other crops gives more yield and income.

- Fallowing of land during the months of November, December and January to make the field hardened.
- Heavy application of farm yard manure to potato every year before ploughing to increase the number of tubers formation.
- Soil amendment with FYM for the reduction of cyst nematode in potato.
- Land stirring by using hand fork to expose the bottom soils to sun.
- Gathering more soil near the stem of the potato crop helpful for the roots to spread over unobstructed and the potatoes were also believed to grow in bigger size.
- Harvest when the leaves of the potato plants turn yellowish brown colour or starts drying after 90 – 100 days of planting.
- Cut the plant stem above the ground level before harvesting of tubers to facilitate the harvesting operation.
- Non-washing of the harvested tubers with water because it may reduce the original colour of the tubers.
- The formation of less than five big sized tubers per plant indicates the failure of potato crop in that particular season.
- The formation of more than 10 big sized tubers per plant indicates the good yield of that particular season.
- Plant of potato eyes towards the slope to prevent water logging and provide good drainage during rainy season.
- Construction of shallow well with 8' depth and 4' width at the lower end of the field to store the draining water for irrigation purposes.

Aromatic crops

Red Oleander

- FYM is abundantly applied during January and August for increased flower yield.
- Oleander up to 5-6 months from planting. Oleander plants are severely pruned to 1-2 feet height at 5-6 years after planting for getting fresh shoots and higher yields. This pruning is

repeated for another 2 times once in 4 years after which they are uprooted and fresh planting is done.

Pepper

- 'Karimundan' is another local variety suitable for higher elevations with short spikes.
- 'Kattumilagu' is a type of wild variety, yielding once in 2-3 years with more pungent berries and it is found in reserve forests only.
- When a sample of pepper is chewed, metallic sound indicates its optimal dryness.
- For producing white pepper, the fruits are allowed to ripen in the climber itself. Then they are collected, put in tanks, foot pressed to remove the skins and they are washed with water and dried to produce white pepper, which is having medicinal value.

Cardamom

- ''Valukkai' producing semi prostrate panicles with smooth surfaced pods is suitable for high rainfall areas.
- Cardamom seeds are sown immediately after harvest to get better germination.
- Seeds, after extraction, are washed well with water for 2 or 3 times to remove the mucilaginous substances, mixed with ash and dried for 2 or 3 days.
- A mixture of neem cake powder and sheep manure is applied @ 200g/plant.
- A mixture of extracts of neem cake and tobacco waste is sprayed to control stem borer and capsule thrips in cardamom.
- Cardamom capsules after the harvest are cured through fumigation. Fumigation is done in a room, locally called as 'store'. There will be 11 trays fitted with sand sieves in a stand and four such stands will be in a store. Capsules are spread on these trays and fumigated. During cardamom curing in the first day morning, the capsules are first preheated at 40° C for 1

hour after which windows are kept open for 1 hour. Then second heating is done at 60° C for 5 hours followed by the opening of windows for 2-3 hours. Then a mild third heating is given for another 10 hours and kept in the trays up to the second day evening. Total curing period takes 30-36 hours. Purpose of curing is to dry the pods without losing their green color to fetch good market value with better keeping quality. Now power base driers are available for processing.

Garlic

- 'Singapore red' is a local variety called as 'periumpoodu' with duration of 4 1/2 months produce bulbs having reddish skin, big sized cloves with less water content and high keeping quality.
- 'Malaippoodu is a local variety with a duration of four months producing big sized bulbs with more pungency and more water content.
- Sheep penning is better for getting higher yields.
- Neem cake is applied 4kg /ac to reduce the infestation for root grub.
- Time of harvest is indicated by yellowing and withering of leaves which turn to pale green colour and start drying from the top. Over maturity causes damage to the bulbs.

Coconut

- To prevent rats from climbing coconut trees, a large palm leaf is split along its mid rib; one set of leaflets is wrapped around the trunk below the crown and the other set is wrapped in the opposite direction.
- To control flower shedding in coconut, salt is poured on the apical portion of the flower buds and also spread at the root zone ad given plenty of water.
- Seed nuts are collected from high yielding mother palms having dense and longer leaves and bigger nuts.
- Seed nuts, which are round in shape and produce metallic sound on tapping are selected for raising nursery.

- Seed nuts are planted in sand bed nursery and kept for six months with irrigation in alternate days.
- About 6-8 months old coconut seedlings having 5-6 leaves are selected for planting
- For coconut planting, pits are dug and filled with Kolingi (*Tephrosia purpurea*) allowed to decompose for six months.
- Suitable seasons for Coconut planting are 'Aadi' (July-Aug.) and Karthigal (Dec.-Jan.) months.
- Before planting coconut seedling, roots are removed in order to induce fresh roots.
- Application of 10 -15kg of FYM per tree every year.
- Application of Kolingi (*Tephrosia purpurea*) @ 10 kg.Aree every year.
- Applying *Calotropis gigantea,* (1kg.), Kolingi (1kg.), Pothakalli (*Poeciloneuron pauciflorum*) (1 kg.) Fishmeal (1 Kg.), salt (1kg.) and sand in a semi-circular basin around the higher yield.
- Mulching by burying of coconut husks around the tree to conserve moisture and to control weeds.
- Inter space in the coconut garden is ploughed twice in a year in Jun.-July and Dec.-Jan. to facilitate aeration to the roots and to control weeds.
- Spraying neem oil to reduce flower shedding.
- To prevent button shedding, common salt is applied around the growing tip @ 2 kg / tree during rainy season.
- Application of ash to control button shedding.
- Kolingi (*Tephrosia purpurea*) and *Calotropis gigantea* are applied in circular basin just before flowering to control button shedding.
- Application of neem cake in the pits before planting coconut, to avoid the attack of insect pests and ants.
- Earthen pots are placed in small pits in coconut gardens and 50% the of the pot is filled with water and ½ kg castor cake.

After three days due to the smell, rhinoceros beetles get attracted, fall in to the pot and die.

- Pouring neem cake extract on the growing tip and adjoining fronds to control rhinoceros beetle.
- To control stem weevil in coconut, the hole bored by it, is cleaned and plugged after putting common salt.
- Putting 1-2 kg of common salt in the pit, while planting coconut, to control termites and to conserve moisture.
- While planting coconut seedling, one leaf of *Agave* spp. is planted in the pit to retain soil moisture and to control termites. Flooding the coconut gaiden to wash off termites
- Lime washing for 2 feet height at the base of coconut trees to control termite attack.
- To control termites, 500 g. of common salt is dissolved in 5 lit. of water and poured on the trunk.
- Grow poultry birds in coconut gardens to feed on termites.
- To control Thanjavur wilt of coconut, green manures like kolingi (*Tephrosia purpurea*), dhaincha (*Sesbania* spp.) etc. are raised and ploughed in situ or well decomposed FYM is applied followed by the application of neem cake.
- To control stem bleeding, the bleeding mouth on the trunk is cut to certain extent, cleaned and poured with lime solution.
- Branches of Seemai karuvel (*Prosopis juliflora*) or barbed wires are tied around the mid trunk to a height of 2-3 feet to prevent climbing of rats and squirrels.
- Greenish yellow coconuts are harvested.
- Adding a piece of jaggery (country sugar) in coconut oil to separate the dusts and make the oil more clear.

Ginger

- Use rhizomes collected near the bunds of the field for seed purpose.
- Application of lime to the field before planting of rhizomes to avoid pests.
- Spreading the available fresh or dried green leaves over the planted rhizomes to avoid weeds and give shading.

Chapter 10

Integrated Organic Farming System

Farming system approach addresses itself to each of the farmer enterprises; inter relationship among enterprises and between the farm and environment. Thus farming system research has the objective of increasing productivity of various enterprises in the farm. Farming system approach introduces a change in farming technique for high production from a farm as a whole with the integration of all the enterprises. The farm produce other than the economic products for which the crop is grown can be better utilized for productive purposes in the farming system approach. A judicious mix of cropping system with associated enterprises like dairy, poultry, piggery, fishery, sericulture etc. suited to the given agro-climatic conditions and socio economic status of farmers would bring prosperity to the farmer.

Combination of Integrated farming system (IFS) along with organic farming so called integrated organic farming system (IOFS)appear to be the possible solution to the continuous increase of demand for food production, stability of income and improvement of nutrition for the small and marginal farmers with limited resources. Integration of different enterprises with crop activity as base will provide ways to recycle products and waste materials of one component as input through another linked component and reduce the cost of production of the products which will finally raise the total income of the farm. This becomes quite essential as crop cultivation is subjected to a high degree of risk and provides only seasonal, irregular and uncertain income and employment to the farmers. With a view to mitigate the risk and uncertainty in agriculture, IOFS serves as an informal insurance.

Production of agricultural crops, vary in response to changes of the seasons. In the recent period stable income of agricultural crops has become unstable. Redressing these by integrating crops with agro-based industries like livestock farming is essential. An integrated organic farming system applies the concept of "Low External Input Sustainable Agriculture" (LEISA) and this system develops the livestock business and the crop business in one location or area using local resources to optimize inputs. Designing a farming system to tie together principles of sustainability and productivity is complex. Organic farmers must consider how the various components of their system - rotations, pest and weed management, and soil health - will maintain both productivity and profitability. This section outlines the major principles incorporated into organic farming systems.

Fig. Components of organic farming system-valley land (0.43 ha)

Efficient cropping systems for a particular farm depend on farm resources, farm enterprises and farm technology because farm is an organized economical unit. The farm resources include land, labour, water, capital and infrastructure. When land is limited intensive cropping is adapted to fully utilized available water and labour when sufficient and cheap labour is available, vegetable crops are also included in the cropping systems as they required more labour. Capital intensive crop like sugarcane, banana, turmeric etc. find a space in the cropping system when capital is not a constraint. In low rainfall regions (750 mm/annum) mono cropping is followed and when rainfall is more than 750 mm, intercropping is practiced, with sufficient irrigation water, triple and quadruple cropping is adopted, when other climatic factors are not limiting farm enterprise like daring, poultry etc. also influenced the type of cropping system. When the farm enterprises include dairy, cropping system should contain fodder crops as components change in cropping system take place with the developments of technology. The feasibility of growing for crop sequences in Genetic alluvial plains inputs to multiple cropping.

Applying an extensive knowledge of indigenous and organic practices, the farm is strategically structured in distinct components that are designed to maximize one another.

A nutrient recycling system generates a virtuous closed loop process on the farm (Figure 1), and biodiversity is intensified to multiply key ecological functions and processes within and among the components (e.g. natural pest management and optimal use of sunlight, rainfall and soil fertility).

Biodiversity-based farming systems are not new. For centuries, farming communities have painstakingly developed resilient and bountiful agricultural systems based on biodiversity, and on their knowledge of how to work with them in equally complex biophysical and socio-cultural settings. Farmers have used diversity for food and economic security through a complex array of home garden designs, agroforestry systems and diversified and integrated lowland farming systems. It differs substantially from conventional modern agriculture in that its focus is the establishment of functional diversity in the farm, rather than monoculture.

The integration of several allied enterprises with crop components is crucial in order to optimize the synergies. These integrated systems provide scope not only to augment income of the farmers but also bring improvement in soil health through recycling of organic wastes and thereby increase the overall productivity of the crops. Thus, energy obtained from IFS in various forms is much higher than energy input, as the by-product/wastes of these allied enterprises provide all raw material and energy required for the food chain in another system. This complimentarity when carefully chosen, keeping in view the soil and environmental conditions generates greater income.

The following strategies should be followed for adoption of integrated organic farming

1. Site Selection/land consolidation

Places which have history of producing crops without using chemical inputs or with minimum intervention, should be preferred

2. Cooperative/community approach

In view of the fragmentation of land-holding, the community approach is a must for the organic farmers.

3. Availability of organic inputs

Easy availability of organic inputs is the pre-requisite for organic farming. The farmers, in due course, have to produce their own organic inputs. The suitability/adaptability of different green manure crops should be tested. All sources of organic material that can (or presently cannot) be used as manure should be identified, this should include industrial wastes also. Gaps in technology that prevent the utilisation of some wastes should then be identified. This should be done to satisfy critics that not enough organic material is available for organic farming.

4. Selection of crops and cultivars

Whether grown for domestic consumption or export purpose.

Selection of crops suited for a particular location.

5. Quality of organic inputs

The organic inputs are sold in different brand names, no standards yet available. Quality control laboratory should be set up to standardize the quality.

6. Cropping system approach

The cropping system approach will be more remunerative in organic farming. Selection of shallow and deep-rooted crops is important in rotation. Part of the crop residue should be returned to soil/fed to cattle or be used for composting.

7. Developmental and promotional activities

Incentive and encouragement for the production of quality organic manure bio-pesticide, bio-fertiliser and green manuring crop should be considered. Effort should be made for the development of new pesticide of plant origin. The uses of bio-agents need to be promoted.

8. Certification and accreditation

Cost of inspection and certification is cost prohibitive. It should be simple and at a lower cost.

9. Sales and marketing

Organic farming is labour intensive. So it will be more remunerative if the farmer gets a premium price for their produce. Promotion of farm level processing, value addition and encouragement of the use of organic farm produce in food industry.

10. Subsidize organic inputs and produce

Subsidies may be provided for organic inputs and produce while the industry is still getting established. In India, subsidies are mainly provided by the national government and channelled through state agriculture departments; the technique is well-tested, having already been used for the synthetic fertilizer and pesticide industry. Indeed, subsidies have been provided for setting up biofertilizer and vermicomposting units under NPOF and for setting up export schemes under NPOP. Additional subsidies could be provided for:

- Setting up organic input production units for composting, biopesticides etc.
- Compensating organic farmers during the period of conversion to organic techniques, to compensate for yield reductions if any.

- Establishing village-level grading and packaging units for organic produce.
- Developing local and regional marketing infrastructure for organic produce in dryland areas, where regional/local food security is more important than crops for export.

11. Develop organic farming clusters of villages.

Since the drylands are already an area of focus for governmental development programs based on a watershed approach, clusters of villages previously established for such programs (Khan, 2002) may be converted into organic clusters of villages by providing technical support. This will be cost-effective and make the eventual certification process of organic produce easier for these villages once the local organic produce market has been well established.

12. Increase public awareness and build capacity.

Conferences, seminars, and farmers' fairs may be organized to raise awareness and encourage adoption of organic farming. Programs demonstrating how to establish organic systems, and training in how to produce and manage organic inputs, may be started at the village level.

Design of integrated farming system model

1. The diversity of the farm should be increased as much as possible by introducing at least 5-6 types of cereals and pulses/ oilseeds, 10-12 varieties of vegetables, 5-6 fruit crops, fuel wood and fodder trees, 5-6 types of spices and medicinal plants, 5-6 livestock, 3-4 types of fish. This could ensure food and livelihood security of the farmer throughout the year.
2. External inputs will have to be reduced. Effective utilisation of resources must be made in the farm to recycle the farm wastes
3. Measures to be taken for conserving the rain water by constructing the water harvesting structures like farm pond and percolation pond.
4. Recycling of farm waste is important.

5. Weeds which are grown in its own on farm, should be processed as a compost and used to meet the consumption requirement of farmer and livestock
6. Establish a manure pit in the corner of the field for composting the farm wastes. Separate for farm wastes and weed materials to be established.
7. Fast growing trees should be planted as they add nutrient to soil and provide habitat for local wildlife, including bird species who also contribute to a healthy ecosystem on the farm.
8. Adjoining land use, buffers
9. Soil fertility management and inputs
10. Proper crop rotation
11. Weed, pest and disease management, materials to be used, and justification
12. Farmers should take initiatives to sell their produce in a processed farm in order to receive more profit. Oil from coconut, groundnut, sesame, fruit juices are few examples of such post-harvest technologies.
13. Integration of livestock at right time and quantum might serve many of our purpose at free of cost. Local breed of ducks in paddy fields, poultry in orchards will save works like weeding fertilizing and aerating the soil.

Before you start your design

Before you start designing your farm, you need to assess your farm according to the following points;

- Existing farm size
- Living area for animal and human
- Ploughing frequency
- Distance of farm aresa from household
- Weeding style and frequency
- Transport after harvest
- Soil water conservation techniques
- Existing farm inputs

- Cropping pattern
- Type of livestock
- Type of fodder

Characteristics of an ideal integrated organic farm

Organic agriculture aims at successfully managing natural resources to satisfy human needs while maintaining the quality of the environment and conserving resources. Organic agriculture thus aims at achieving economic, ecological and social goals at the same time:

1. Ecological goal: "How does the farm improve nature and survival of other organisms?"
2. Social goal: "How do other people benefit from the farm?"
3. Economic goal: "What benefits do I generate from the farm?"

The ecological goal

The ecological goal basically relates to maintenance of quantity and quality of natural resources. Farming should be done in an environmentally-friendly manner, whereby the soil, water, air, plants and animals are protected and enhanced. Organic farmers pay special attention to the fertility of the soil, the maintenance of a wide diversity of plants and animals, and to animal friendly husbandry.

Important environmental goals are:

- Prevention of loss and destruction of soil due to erosion and compaction.
- Increasing the humus content of the soil.
- Recycling farm-own organic materials and minimizing use of external inputs.
- Promotion of natural diversity of organisms - being a criterion of a balanced natural ecosystem.
- Prevention of pollution of soil, water and air through avoidance of fertilisers and pesticides.
- Ensuring husbandry that considers natural behaviour of farm animals.
- Use of renewable energy, wherever possible.

To achieve these goals organic farmers maintain wide crop rotations, practice intercropping and cover cropping, plant hedgerows and establish agro-forestry systems.

The social goal

Organic farming aims at improving the social benefits to the farmer, his/her family and the community in general.

Important social goals include

- Creating good working conditions for all.
- Ensuring a safe nutrition of the family with healthy foods.
- Ensuring sufficient production for subsistence and income.
- Encouraging fair and conducive working conditions for hired workers.
- Encouraging learning and application of local knowledge.

From an organic perspective, at the household level fair participation in farm activities of all family members and proper sharing of the benefits from the farm activities is essential. On community level, knowledge and experiences should be shared, and collaboration strengthened in order to obtain higher benefits.

The economic goal

In an economic sense organic farming aims at optimizing financial benefits to ensure short- and long-term survival and development of the farm. An organic farm should not only pay for production costs, but also meet the household needs of the farmer's family.

Important economic goals include

- Satisfactory and reliable yield.
- Low expenditures on external inputs and investments.
- Diversified sources of income for high income safety.
- High value added on-farm products through improvement of quality and on-farm processing of products.
- High efficiency in production to ensure competitiveness.

Organic farmers try to achieve this goal by creating different sources of income from on- and off-farm activities. Usually different crop and animal enterprises are adopted simultaneously in a mixed production system. The target also includes being more self-sufficient in terms of seeds, manures, pesticides, food, feeds, and energy sources and thereby minimizing cash outlay to purchase off-farm items.

Strategies to improve long-term productivity of the integrated organic farm

Reduce production risks

- Diversification
- Build soil fertility
- Reduce external inputs

Improved overall production

- Use improved adapted local varieties
- Improve soil fertility
- Ensure proper pest and disease management
- Integrate livestock

Enhance value of farm products

- Adopt profitable enterprises
- Improve product quality
- Establish storage and processing facilities
- Obtain organic certification

Reduce expenses

- Reduce own manure
- Produce own planting materials and seeds
- Make own herbal pesticides and organic inputs
- Share equipment and machinery

Integrated organic Farm Model using 2.5 acres (1.0 ha) land

Following is a model (Figure 2) which could be used in garden land area of Tamil Nadu. This model comprises the following subsystems:

1. Crops production (grains, root crops, coconut, fruit trees, vegetables) - 8250 m^2
2. Fodder crops - 1000 m^2
3. Livestock – Cattle, poultry, Goat in 100 m^2
4. Biodigester - 20 m^2
5. Compost/vermiculture - 60 m^2
6. Pest repellent cafeteria - 100 m^2
7. Organic fertilizer production -20 m^2
8. Area for proper land use – 100 m^2
9. Rain water harvesting – 100 m^2
10. Bee hives – 20 m^2
11. Agroforestry - 200 m^2
12. Kitchen garden – 30 m^2

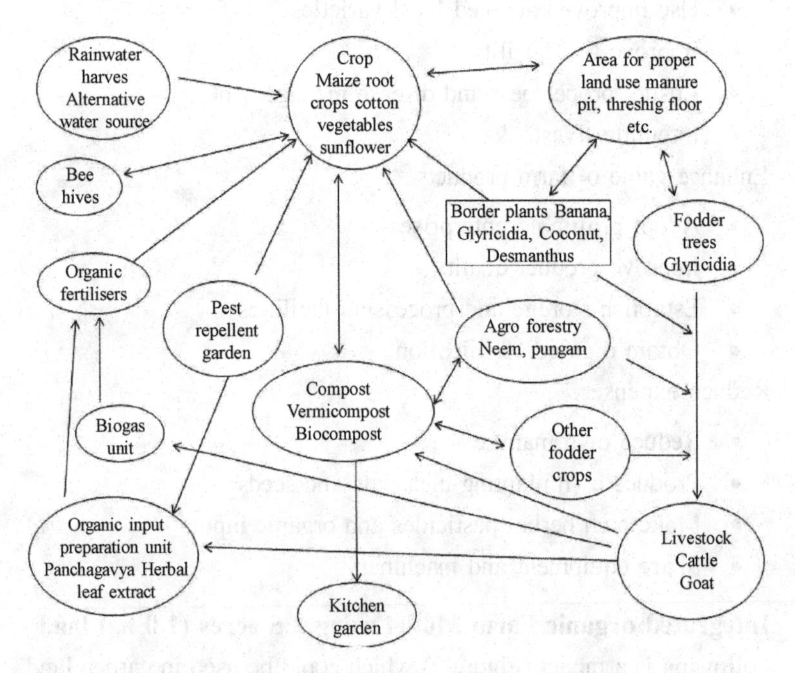

Fig. 2. Integrated organic farming system model

Table 17. Components of organic farming system model of TNAU (Irrigated dryland ecosystem)

Components	Treatments/ Remarks
Crop component	Cropping Systems, 1. Bhendi + Leaf coriander - Maize + Cowpea (Fodder), 2. Green manure - Cotton - Redgram, 3. Fodder grass and Desmanthus
Agro forestry	*Azardhiracta indica, Melia dubia, Sesbania grandiflora, Pongamia pinnata, Gmelina arborea, Ailanthus excelsa, Cajanus cajan* and *Sesbania sesban*
Dairy	Milch animal: 2 cows with one calf
Vermicompost unit	The residue of the crops and manure from the dairy unit will be converted into vermicompost and recycled as manure for crops
Area under supporting activities	Cattle shed, manure pit, threshing floor etc.
Border plants	Annual Moringa,, Coconut, Banana, *Hedge lucerne Desmanthus.* Curry leaf, Coriander Mintefe etc.,

Dairy unit | Livestock | Desmanthus in the border

Live fence | Kitchen garden | Biomass - Dhanicha

Manure pit | Vermicompost | Glyricidia fencing

Fig. TNAU IOFS model

Challenges

A key challenge of such a multi-faceted system is the diversity of the skills required, particularly following natural disasters, where different things need to be fixed. On the other hand, by diversifying their skills, the farmers empowered themselves and improved their self-confidence. Meanwhile, it is important to recognize that the stable access to natural resources, land and water on the farm facilitates its success. A fool proof winning attitude from the family farmers definitely appears to be another critical factor for success. The major obstacles in practicing pure organic agriculture have been identified as limited technological options, large marginal costs and risk in shifting to a new system from the conventional farming, low awareness about the organic farming system, lack of marketing and technical infrastructure and added cost by way of inspection, certification.

Benefits of integrated organic farming system

- *Productivity:* IOFS provides an opportunity to increase economic yield per unit area per unit time by virtue of intensification of crop and allied enterprises.
- *Profitability*: Use waste material of one component at the least cost. Thus reduction of cost of production and form the linkage of utilization of waste material and elimination of middleman interference in most inputs used. Working out net profit/ BC ratio is increased.
- *Potentiality or Sustainability*: Organic supplementation through effective utilization of byproducts of linked component is done thus providing an opportunity to sustain the potentiality of production base for much longer periods.
- *Balanced Food*: Components of varied nature are linked to produce different sources of nutrition. Environmental Safety: In IOFS waste materials are effectively recycled by linking appropriate components, thus minimize environment pollution.
- *Recycling*: Effective recycling of waste material (crop residues and livestock wastes) in IOFS. Therefore, there is less reliance to outside inputs – fertilizers, agrochemicals, feeds, energy, etc.

- *Income Rounds the year*: Due to interaction of enterprises with crops, eggs, milk, mushroom, honey, cocoons silkworm, it provides flow of money to the farmer round the year. There is higher net return to land and labour resources of the farming family.
- *Adoption of New Technology*: Resourceful farmers (big farmer) fully utilize technology. IOFS farmers, linkage of dairy/ mushroom / sericulture / vegetable. Money flow round the year gives an inducement to the small/ original farmers to go for the adoption of technologies.
- *Saving Energy*: To identify an alternative source to reduce our dependence on fossil energy source within short time. Effective recycling technique the organic wastes available in the system can be utilized to generate biogas. Energy crisis can be postponed to the later period.
- *Meeting Fodder crisis:* Every piece of land area is effectively utilized. Plantation of perennial legume fodder trees on field borders and also fixing the atmospheric nitrogen. These practices will greatly relieve the problem of non – availability of quality fodder to the animal component linked.
- *Solving Fuel and Timber Crisis*: Linking agro- forestry appropriately the production level of fuel and industrial wood can be enhanced without determining effect on crop. This will also greatly reduce deforestation, preserving our natural ecosystem.
- *Employment Generation*: Combing crop with livestock enterprises would increase the labour requirement significantly and would help in reducing the problems of under employment to a great extent. IOFS provide enough scope to employ family labour round the year.
- *Agro – industries*: When one of produce linked in IOFS are increased to commercial level there is surplus value adoption leading to development of allied agro – industries.
- *Increasing Input Efficiency*: IOFS provide good scope to use inputs in different component greater efficiency and benefit cost ratio.

Usefulness of Organic Farming in the context of System Approach

- Organic manures improves soils physico-chemical and biological properties and produces optimal condition in the soil for high yields and good quality crops
- Reduces cost of purchased inputs
- Farm wastes and residues are effectively recycled thus reducing environmental pollution and can be used to regenerate degraded areas.
- Organic farming allows the promotion of biodiversity which is vital for ecological balance
- Helps to prevent environmental degradation.
- Increases soil organic carbon
- Enhances soil microbial population
- Carbon sequestration and stock in soil

Organic farming in the cropping system perspective will be a viable avocation to address the sustainability aspects. Effective input management, water and plant protection is possible if organic production techniques are followed on system basis. Restoration of soil fertility under intensive organic farming situations could be possible through the inclusion of legumes/green manures in the cropping system. Under certified organic agriculture, cropping system based production strategies alone will minimize the cost of production and maximize the net profit through optimum resource utilization. Sustainable organic agriculture relies on resource conservation, which is best possible under cropping system mode. Integrated farming systems offer unique opportunities for maintaining and extending biodiversity. The emphasis in such systems is on optimizing resource utilization rather than maximization of individual elements in the system. The wellbeing of poor farmers can be improved by bringing together the experiences and efforts of farmers, scientists, researchers, and students in different countries with similar eco-sociological circumstances i.e. through Integrated Organic Farming System (IOFS).

Chapter 11

Organic Certification and Legislation of Organic Food

In general, any business directly involved in food production can be certified, including seed suppliers, farmers, food processors, retailers and restaurants. Requirements vary from country to country, and generally involve a set of production standards for growing, storage, processing, packaging and shipping organic certification ensures:

- Avoidance of synthetic chemical inputs (e.g. Fertilizer, pesticides, antibiotics, food additives, etc) and genetically modified organisms;
- Use of farmland that has been free from chemicals for a number of years (often, three or more);
- Keeping detailed written production and sales records (audit trail);
- Maintaining strict physical separation of organic products from non-certified products and
- Undergoing periodical on-site inspections.

In some countries, certification is overseen by the government, and commercial use of the term *organic* is legally restricted. Certified organic producers are also subject to the same agricultural, food safety and other government regulations that apply to non-certified producers.

Purpose of certification

Organic certification addresses a growing worldwide demand for organic food. It is intended to assure quality and prevent fraud. For organic producers, certification identifies suppliers of products approved for use in certified operations. For consumers, "certified organic" serves as a product assurance, similar to "low fat", "100% whole wheat", or "no artificial preservatives". Certification is essentially aimed at regulating and facilitating the sale of organic products to consumers. Individual certification bodies have their own service marks, which can act as branding to consumers—a certifier may promote the high consumer recognition value of its logo as a marketing advantage to producers. Most certification bodies operate organic standards that meet the national government's minimum requirements.

The certification process

In order to certify a farm, the farmer is typically required to engage in a number of new activities, in addition to normal farming operations:

Study the organic standards, which cover in specific detail what is and is not allowed for every aspect of farming, including storage, transport and sale.

Compliance: farm facilities and production methods must comply with the standards, which may involve modifying facilities, sourcing and changing suppliers, etc.

Documentation: extensive paperwork is required, detailing farm history and current set-up, and usually including results of soil and water tests.

Planning: a written annual production plan must be submitted, detailing everything from seed to sale: seed sources, field and crop locations, fertilization and pest control activities, harvest methods, storage locations, etc.

Inspection: annual on-farm inspections are required, with a physical tour, examination of records, and an oral interview.

Fee: A fee is to be paid by the grower to the certification body for annual surveillance and for facilitating a mark which is acceptable in the market as symbol of quality.

Record-keeping: written, day-to-day farming and marketing records, covering all activities, must be available for inspection at any time. In addition, short-notice or surprise inspections can be made, and specific tests (e.g. soil, water, plant tissue) may be requested. For first-time farm certification, the soil must meet basic requirements of being free from use of prohibited substances (synthetic chemicals, etc) for a number of years. A conventional farm must adhere to organic standards for this period, often, three years. This is known as being in *transition*. Transitional crops are not considered fully organic. A farm already growing without chemicals may be certified without this delay.

In India, there are two accreditation systems for authorizing Certification and Inspection agencies for organic certification. National Programme on organic Production (NPOP) promoted by Ministry of Commerce is the core programme which governs and defines the standards and implementing procedures. National Accreditation Body (NAB) is the apex decision making body. Certification and Inspection agencies accredited by NAB are authorized to undertake certification process. The NPOP notified under FTDR act and controlled by Agricultural Processed Foods Export Development Authority (APEDA) looks after the requirement of export while NPOP notified under APGMC act and controlled by Agriculture Marketing Advisor, Directorate of Marketing and Inspection looks after domestic certification. Currently 26 certification agencies have been authorized to undertake certification process. This type of certification is mostly referred as third party certification.

National Programme on Organic Production

National Program on Organic Production (NPOP) was launched during 2001 under the Foreign Trade & Development Act (FTDR Act). The document provides information on standards for organic production, systems criteria, and procedures for accreditation of Inspection and Certification bodies, the national organic logo and the regulations governing its use.

National Standards for Organic Production (NSOP)

National Standards for Organic Production are grouped under following six categories:

1) Conversion
2) Crop production
3) Animal husbandry
4) Food processing and handling
5) Labeling
6) Storage and transport

Standard requirements for crop production, food processing and handling are listed below:

Conversion Requirements

The time between the start of organic management and cultivation of crops or animal husbandry is known as the conversion period. All standard requirements should be met during conversion period. Full conversion period is not required where organic farming practices are already in use.

Crop production

Choice of crops and varieties – All seeds and planting materials should be certified organic. If certified organic seed or planting material is not available then chemically untreated conventional material can be used. Uses of genetically engineered seeds, pollen, transgenic plants are not allowed.

Duration of conversion period – The minimum conversion period for plant products, produced annually is 12 months prior to the start of the production cycle. For perennial plants (excluding pastures and meadows) the conversion period is 24 months from the date of starting organic management. Depending upon the past use of the land and ecological situations, the certification agency can extend or reduce the minimum conversion period.

Fertilization policy: Biodegradable material of plant or animal origin produced on organic farms should form the basis of the fertilization policy. Fertilization management should minimize nutrient losses, avoid accumulation of heavy metals and maintain the soil pH. Emphasis should be given to generate and use own on farm organic fertilizers. Brought in fertilizers of biological origin should be supplementary and not a replacement. Over manuring should be avoided. Manures containing human excreta should not be used on vegetation for human consumption.

Pest disease and weed management including growth regulators: Weeds, pests and diseases should be controlled preferably by preventive cultural techniques. Botanical pesticides prepared at farm from local plants, animals and microorganisms are allowed. Use of synthetic chemicals such as fungicides, insecticides, herbicides, synthetic growth regulators and dyes are prohibited. Use of genetically engineered organisms or products is prohibited. All equipments from conventional farming systems shall be properly cleaned and free from residues before being used on organically managed areas.

Soil and water conservation: Soil and water resources should be handled in a sustainable manner to avoid erosion, salinisation, excessive and improper use of water and the pollution of surface and ground water. Cleaning of land by burning (e.g. slash and burn and straw burning) should be restricted. Clearing of primary forest for agriculture (jhuming or shifting cultivation) is strictly prohibited.

Collection of non-cultivated material of plant origin and honey: Wild harvested products shall only be certified organic, if derived from a stable and sustainable growth environment and the harvesting shall not exceed the sustainable yield of the ecosystem and should not threaten the existence of plant or animal species. The collection area should not be exposed to prohibited substances and should be at an appropriate distance from conventional farming, human habitation, and places of pollution and contamination.

Food processing and handling

General principles: Organic products shall be protected from co-mingling with nonorganic products, and shall be adequately identified through the whole process. Certification programme shall regulate the means and measures to be allowed or recommended for decontamination, clearing or disinfection of all facilities where organic products are kept, handled, processed or stored. Besides storage at ambient temperature the following special conditions of storage are permitted.

- 100% of the ingredients of agricultural origin shall be certified organic
- Water and salt may be used as organic products
- Preparations of microorganisms and enzymes commonly used in food processing may be used with the exceptions of genetically engineered microorganisms and their products
- Co-mingling with inorganic products shall be prevented
- Proper storage under controlled atmosphere, cooling, Freezing drying and humidity regulation is needed
- Mechanical, physical and biological methods to control pests only recommended
- Carcinogenic pesticides and disinfectants are not permitted
- Use of minerals and vitamins as additives and processing aids shall be restricted
- Processing by mechanical, smoking, extraction, precipitation, filtration only approved
- Irradiation is not allowed
- Eco-friendly, biodegradable packaging materials is to be used
- Product integrity should be maintained
- Proper labeling of products is must

Controlled atmosphere, cooling, freezing, drying and humidity regulation.

Pest and disease control: For pest management and control following measures shall be used in order of priority Preventive methods such as disruption, and elimination of habitat and access to facilities. Other methods of pest control are:

- Mechanical, physical and biological methods
- Permitted pesticidal substances as per the standards and

- Other substances used in traps.
- Irradiation is prohibited.
- Direct or indirect contact between organic products and prohibited substances (such as pesticides) should not be there.

Packaging

Material used for packaging shall be ecofriendly. Unnecessary packaging material should be avoided. Recycling and reusable systems should be used. Packaging material should be biodegradable. Material used for packaging shall not contaminate the food.

Labelling

When the full standard requirements are met, the product can be sold as "Organic". On proper certification by certification agency "India Organic" logo can also be used on the product.

Storage and transport

Products integrity should be maintained during storage and transportation of organic products. Organic products must be protected from co-mingling with non-organic products and must be protected all times from contact with the materials and substances not permitted for use in organic farming.

General requirement for certification

1. A registered operator shall Comply with National Programme for Organic Production (NPOP) norms and shall adhere to the National Standards for Organic Production (NSOP) and TNOCD general standards for organic agricultural production, animal husbandry production, honey, wild collection, processing, packaging, storage, labelling and transport standards.
2. Prepare, implement, and update annually an organic production plan and submit to Tamil Nadu Organic Certification Department (TNOCD) every year.
3. Permit on-site inspections with complete access to the production and handling operation, including non certified

production and handling operation, areas, structures, offices by the Organic Certification Inspectors and other higher officials of TNOCD and also officials of APEDA whenever required.

4. Maintain all records applicable to the organic operation for not less than 5 years after creation of such records and allow authorized representatives of TNOCD, State or Central Government officials of accrediting agency access to such records during normal working hours for review and copying to determine compliance with NPOP norms and TNOCD Standards.
5. Pay the prescribed fees charged by TNOCD within stipulated time.
6. Operator shall inform the TNOCD in case of any
   a. Application, including drift, of a prohibited substances to any, production unit, site, facility, livestock, or product that is part of an operation and
   b. Changes in certified operations or any portion of a certified operation that may affect the organic integrity in compliance with standards of NPOP and TNOCD.

Application for certification

A person seeking organic certification of production or handling operation shall submit application for registration in the prescribed format in triplicate. The application shall include the following information

1. An organic production or handling system plan,
2. All information requested in the application shall be completed in full i.e. name, addresses, details of contact person, telephone number of the authorized person etc.,
3. The names of organic certification body to which application is previously made and out come, non-compliance noted if any, copy of such records and reason for applying shall be given.
4. Any other information necessary to determine the compliance with the standards specified.

5. The prescribed registration fee, one time inspection fee, one time travel cost shall be paid by the operator along with the application form. The other prescribed fees shall be paid by the operator as notified by TNOCD during the course of certification process.

Review of application

1. Application shall be scrutinized.
2. Any information required shall be communicated to the operator and operator shall submit the requested information immediately.
3. Application without prescribed fee shall not be reviewed.
4. After review of application decision shall be made by TNOCD on acceptance/ rejection of the application.
5. The rejected application shall be returned to the applicant citing reasons for rejection along with the fees enclosed.
6. Fee paid for the applications accepted by TNOCD shall not be refunded at any circumstances.
7. An initial onsite inspection shall be fixed and communicated to the operator after registration or shall be noted in the registered copy of application itself.
8. An applicant can withdraw the application at any time but the fees paid shall not be refunded.

Scheduling of inspection

1. Initial field inspection shall be fixed at a reasonable time so that the operator can demonstrate compliance or capacity to comply with the standards while conducting inspection of land, facilities and activities. Such initial onsite inspection shall be delayed up to six months from the date of registration so as to give time for the operator to comply with required standards including record keeping.
2. All onsite inspection shall be conducted only in the presence of operator or an authorized representative of the operator who is knowledgeable about the operation. However this

requirement does not arise in the case of unannounced / surprise inspections.

3. There shall be one annual inspection and additional inspection shall be fixed based on the risk assessment carried out during initial inspection.

Verification during inspection

1. During the field inspection, the OCI shall verify the compliance or the capacity to comply with the NPOP standards and TNOCD standards.
2. Verification of information on organic production plan submitted by the operator and practical implementation of the standards.
3. OCI shall ensure that the prohibited substances/ materials are not used and in case of suspicion the OCI, shall draw samples of soil, water, wastes, seeds, plant tissues, plant, animal and processed products.
4. The samples shall be tested in NABL accredited ISO 17025 laboratories. The operator shall bear the cost of samples sent for analysis.
5. During onsite inspection the OCI shall conduct interview with the person responsible for the organic production system to confirm accuracy of information gathered during inspection and completeness of inspection, observation gathered during the onsite inspection. The inspector shall also collect other required information as well as issues of concern.
6. After inspection the OCI shall prepare checklist and inspection report and obtain signature of the operator or his representative.
7. A copy of the check list and inspection report shall be sent to the concerned operator and Evaluator.
8. Inspection reports shall be evaluated by the evaluator within reasonable time and any additional information required shall be addressed to the operator.
9. In case of any non compliance to the prescribed standards an explanation shall be called from the operator and sanctions shall be imposed if required.

Continuation of certification

1. To continue certification the operator shall renew registration by paying fees for renewal.
2. An updated annual report for production or handling operation shall be submitted by the operator.
3. An updated corrective action for minor non conformities previously identified shall be submitted by the operator.
4. TNOCD after receipt of renewal application for continuation of certification shall scrutinize the application and verify the facts.

Fair trade

All the operators shall perform their operation with social justice; they shall not employ child labour, and shall protect rights of women, smallholder, traditional agriculture and indigenous people's rights.

Appeal

1. Registered operator may appeal against the notice of denial of certification, proposed suspension or revocation to the appellate authority (Director, TNOCD).
2. An appeal shall be made within the time period mentioned in the notification or within 30 days from the date of receipt of the notification, whichever occurs later. The appeal shall be considered filed on the date of receipt in the office of Director, TNOCD. The decision of the appellate authority shall be final.

Initiating organic farming

Dry lands are the potential places where organic farming can be started because in dry lands

1. Less effect of high input agriculture thus least residue of pesticide and less time required for conversion.
2. Organic manures improve the fertility and water retention capacity of poor soils of dry lands.
3. Poor economic status of the dryland farmers limits them to purchase high cost inputs. But they can do the labour intensive operations to support organic farming.

Action plan for promotion of organic farming

The salient features of crop production management in organic farming include use of organic manures, recycling of organic wastes, proper crop rotation, intercropping, mixed cropping and poly-cropping, green manure cropping, use of biofertilizers, mulching of weeds, integrated pest management, judicious use of irrigation water. The action plan is follows.

1. Development of organic farming practices

Model organic farms can be developed for the crops amenable for organic farming and package of practices can be developed for further dissemination and adoption.

2. Imparting training on organic farming

The small and medium farmers, NGOs and private entrepreneurs have to be trained on the best organic farming practices, storage and processing of organic products. The benefits of organic farming can be elaborated to the participants of such training programmes. Demonstration in selected locations for specific crops can be conducted in the farmers' fields to enable the farmers to gain acquaintance with new technology packages.

3. Identification of areas and villages for organic farming

Organic farming can be promoted for a suitable region or a village or cluster of farms to derive maximum benefits of organic farming. This will also facilitate in transport and marketing of the crop produces. Organic farming societies / farmers' forums can be established at village levels. The forums can help the farmers in getting soft loans from banks and registration in the institute identified by the government for organic certification.

4. Facility creation for organic farmers

Facilities should be created for storage, processing, packaging, transport, quality control and sales mechanisms at taluk /district levels and to ensure premium price for the organic farm produces. To begin with agricultural marketing department can be entrusted for facility creation activities at district level and then to taluk and block levels. The market intelligence cells can be created at State Agricultural Universities which can co-ordinate with Government agencies, NGOs, farmers'forums in assessing the market rates, market facilities at national and international levels. Basic training on food processing, post harvest technology and market intelligence can be imparted to different stakeholders at the State Agricultural Universities.

5. Organic certification

The growing awareness among the public and farmers on the ill effects of chemical agriculture and increasing demand for the safe and quality food necessitated the promotion of organic agriculture in India. Since, organic certification improves the image of organic agriculture and provides transparency in certification, Government agencies can be assigned with organic certification program. The certification facility also can be utilized by different stake holders including local bodies, NGOs, private entrepreneurs and self help groups (SHGs) who produce and market organic commodities in large quantities.

Participatory Guarantee System

Participatory Guarantee System (PGS) is a quality assurance initiative that is locally relevant, emphasize the participation of stakeholders, including producers and consumers and operate outside the frame of third party certification. As per IFOAM (2008) definition "Participatory Guarantee Systems are locally focused quality assurance systems. They certify producers based on active participation of stakeholders and are built on a foundation of trust, social networks and knowledge exchange". PGS is a process in which people in similar situations (in this case small holder producers) assess, inspect and verify the production practices of each other and collectively declare the entire holding of the group as organic.

PGS system has number of basic elements which embrace a participatory approach, a shared vision, transparency and trust. Participation is an essential and dynamic part of PGS. Key stakeholders (producers, consumers, retailers and traders and others such as NGOs) are engaged in the initial design, and then in the operation of the PGS. In the operation of a PGS, stakeholders (including producers) are involved in decision making and essential decisions about the operation of the PGS itself. In addition to being involved in the mechanics of the PGS, stakeholders, particularly the producers are engaged in a structured ongoing learning process, which helps them improve what they do. This process is facilitated by the PGS group itself or in some situations a supportive NGO. The learning process is usually 'hands-on' and involves field days or workshops. The idea of participation 4 embodies the principle of collective responsibility for ensuring the organic integrity of the PGS.

Guiding Principles for Organic Participatory Guarantee System In tune with the international trends and IFOAM's PGS Guidelines,

PGS India system is also based on participatory approach, a shared vision, transparency and trust. In addition it gives PGS movement a National recognition and institutional structure without affecting the spirit of PGS.

1. Participation

Participation is an essential and dynamic part of PGS. Key stakeholders (producers, consumers, retailers, traders and others such as NGOs) are engaged in the initial design, and then in the operation of the PGS and decision making. The idea of participation embodies the principle of a collective responsibility for ensuring the organic integrity of the PGS. This collective responsibility is reflected through:

- Shared ownership of the PGS
- Stakeholder engagement in the development process
- Understanding of how the system works and
- Direct communication between producers and consumers and other stakeholders Together these help shape the integrity based approach and a formula for trust. An important tool for promoting this trust is having operational processes that are transparent. This includes transparency in decision making, easy access to the data base and where possible farms are open to participation and visits of consumers. Participation of traders/

retailers or consumers in decision making may not be possible under all situations, but their participation in any form will increase the credibility and trustworthiness of the group.

2. Shared Vision

Collective responsibility for implementation and decision making is driven by common shared vision. All the key stakeholders (producers, facilitating agencies, NGOs, social organizations and even the State Governments) support the guiding principles and goals, PGS is striving to achieve. This can be achieved initially through their participation and support in the design and then by joining it. This may include commitment in writing through signing an application/ document that includes the vision. Each stakeholder organization (or PGS group) can adopt its own vision conforming to the overall vision and standards of PGS India

3. Transparency

Transparency is created by having all stakeholders, including producers and consumers, aware of exactly how the guarantee system works to include the standards, the organic guarantee process (norms) with clearly defined and documented systems and how decisions are made. Public 5 access will be ensured to documentation and information about the PGS groups, such as lists of certified producers and details about their farms and non-compliance actions. These will be available through a dedicated National database websites. But still it does not mean that entire information on National PGS database will be available to everyone. At the grass roots level transparency is maintained through the active participation of the producers in the organic guarantee process which can include • Information sharing at meetings and workshops • Participation in internal inspections (peer reviews) • Involvement in decision making.

4. Trust

The integrity base upon which PGS are built is rooted in the idea that producers can be trusted and that the organic guarantee system can be an expression and verification of this trust. The foundation of this trust is built from the idea that the key stakeholders collectively develop their shared vision and then collectively continue to shape and reinforce their vision through the PGS. The ways this trust is reflected may depend entirely on factors that are culturally/ socially specific to the PGS group. The idea of 'trust' assumes that the individual producer has a commitment to protecting nature and consumers' health through organic production. Mechanism for expressing trustworthiness includes: • Declaration (a producer pledge) via a witnessed signing of a pledge document • Written collective undertaking by the group to abide by the norms, principles and standards of PGS.

5. Horizontality

PGS India is intended to be non-hierarchical at group level. This will reflect in the overall democratic structure and through the collective responsibility of the PGS group with sharing and rotating responsibility, by engaging producers directly in the peer review of each other's farms; and by transparency in decision making process.

National networking

PGS India while keeping the spirit of PGS intact also aims to give the entire movement an institutional structure. This is proposed to be achieved by networking the groups under common umbrella through various facilitating agencies, Regional Councils and Zonal Councils. To make the system completely transparent and accessible to traders and consumers entire data will be hosted on a common platform in the form of a website. National Centre of Organic Farming shall be the custodian of data, define policies and guidelines and undertake surveillance through field monitoring and product testing for residues. Regional councils and facilitating 6 agencies will facilitate the groups in capacity building, training, knowledge/ technology dissemination and data uploading on the PGS website. But at every stage it will be ensured that these agencies including apex body do not interfere in the working and decision making of the group. Even if surveillance is done and reports are made, the same will also be put on website in public domain. What action is to be taken on adverse reports will be left to the group and regional council.

Advantages of PGS over third party certification system

In PGS organic farmers have full control over the certification process and are able to produce far more credible and effective system of quality assurance compared to third party certification. Important benefits of this system over third party certification system are as follows:

- The procedures are simple; documents are basic and use the local language be understandable to farmers.
- All the members are local and known to each other. Being themselves practicing organic farmers have high degree of understanding on dayto-day knowledge or acquaintance of the farm.
- Peer appraisers are among the group and live in the same village, therefore have better access to surveillance
- Peer appraisal instead of third party inspections reduces cost
- Mutual recognition and support between Regional PGS groups ensures better networking for processing and marketing.
- Empowers farmers with increased capacity building. Bring consumers to the farm without the need of middleman.
- Unlike grower group certification system, PGS offer every farmer with individual certificate and each farmer is free to market its own produce independent of group.
- Consumers and buyers are often involved in production and verification process
- Random residue testing at regular intervals ensures the integrity and increases the trust.

Limitations of PGS

PGS certification is only for farmers or communities that can organize and perform as a group within the village or in close-by villages with continuous territory and is applicable on, on-farm activities comprising of crop production, processing and livestock rearing (including bee keeping) and off-farm processing "by PGS farmers of their direct products".

Individual farmers or group of farmers having less than 5 members are not covered under PGS. They either have to opt for third party certification or join the existing PGS local group.

PGS is applicable on on-farm activities comprising of crop production, processing and livestock rearing and off-farm processing "by PGS farmers of their direct products". Off-farm processing activities such as, storage, transport and value addition activities by persons/agencies other then PGS farmers away from the group are not covered under PGS. Off-farm input approval granted by the group is applicable on the members of the same group and cannot be taken as a basis for universal approval for other groups. Off-farm inputs need to be approved by each group for their member's use on case to case basis.

PGS ensures traceability only up to end till it is in the custody of PGS group. Once the product leaves the custody of PGS group there is no control of PGS on its integrity, Therefore PGS is ideal for local direct sales/ direct trade between producer and consumer and direct trade of packed finished product with PGS logo between PGS group and traders/ retailers. But Local Groups and buyers in consultation with RC can devise some mechanism with full traceability records to allow use of PGS logo on products packed by traders/ retailers.

Operational Structure

Schematic operational structure of the PGS India is given below:

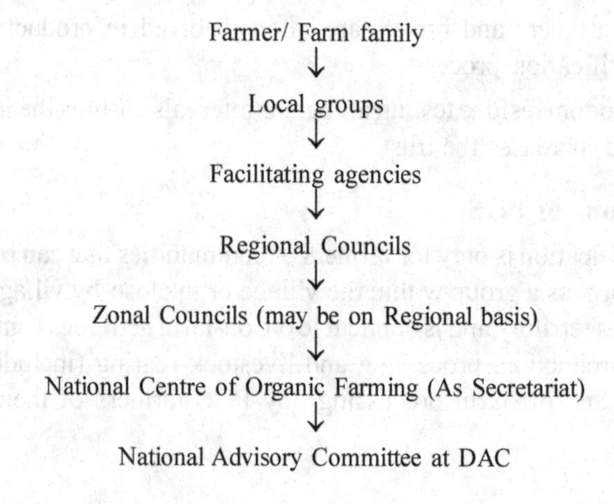

Organic Food Regulations in India

Consumer's interest is growing towards organic foods is increasing due to food safety and environment concerns. However, consumers are reluctant to buy organic foods due to plethora of organic foods available in the market and they are not sure about their genuineness. To ensure credibility of organic foods, there is a need for a regulatory mechanism backed by credible certification system.

The two systems of certification i.e. National Programme for Organic Production (NPOP) implemented by APEDA in the Department of Commerce and Participatory Guarantee System for India (PGS-India) implemented by National Centre for Organic Farming (NCOF) in the Ministry of Agriculture and Farmers Welfare are being popularly implemented in the country.

I. National Programme for Organic Production (NPOP): NPOP launched during 2001 is a quality assurance initiative by the Government of India under the Ministry of Commerce and Industry. The NPOP not only provides the institutional framework for accreditation of certification agencies and operationalization of certification programme through its accredited certification agencies but also ensures that the system effectively works and is monitored on regular basis. 29 Accredited certification agencies authorized under the programme are certifying organic producers.

Scope of NPOP- It covers Crop production and wild harvest, Livestock, Sericulture, Apiculture, Aquaculture, Organic Food processing and Handling, Organic Animal Feed Processing and Handling, Organic Mushroom and Seaweeds, Aquatic Plants and Green House Crop Production.

Objectives

a. To provide the means of evaluation of certification programme for organic agriculture and products as per the approved criteria.

b. To accredit certification programmes of Certification Bodies seeking accreditation under NPOP.

c. To facilitate certification of organic products in conformity with the National Standards for Organic Production (NSOP).

d. To facilitate certification of organic products in conformity with the importing countries organic standards as per equivalence agreement between the two countries or as per importing country requirements.

e. To encourage the development of organic farming and organic processing.

India Organic Logo- Certification mark of NPOP

II. Participatory Guarantee System for India (PGS-India) - PGS is a decentralized and farmers group centric quality assurance initiative built up by locally known farmers in which farmers group among themselves assess, inspect and verify compliance of PGS-India standards for production practices collectively to declare the entire land holding of the group as organic. Participatory Guarantee Systems certify producers based on active participation of stakeholders and are built on a foundation of trust, social networks and knowledge exchange. People in similar situations (small holder producers) assess, inspect and verify the production practices of eachother and take decision on organic certification. National Advisory Committee is the apex policy making body for PGS India Programme. National Centre of Organic Farming is the Secretariat of the PGS programme with Director NCOF as the Executive authority.

- **Scope of PGS-India-** It covers crop production, wild harvest, aquaculture, livestock, beekeeping, food processing and handling.

PGS-India Organic Logo- Certification mark of PGS-India

PGS-INDIA ORGANIC

Food Safety and Standards (Organic Foods) Regulations, 2017

Section 22 of the Food Safety Standards Act, 2006 (FSS Act, 2006) gives mandate to Food Safety and Standards Authority of India (FSSAI)to regulate manufacture, distribute, sell or import organic foods. Accordingly, FSSAI has notified the Food Safety and Standards (Organic Foods) Regulations, 2017 under the FSS Act, 2006.

Salient Features of the Food Safety and Standards (Organic Foods) Regulations, 2017:

i. **Mandatory Requirement** - No person shall manufacture, pack, sell, offer for sale, market or otherwise distribute or import any organic food unless they comply with the requirements laid down under these regulations.

ii. **Applicability of the systems-** The organic food offered or promoted for sale shall comply with all the applicable provisions of one of the following systems, namely:—

a. National Programme for Organic Production (NPOP)

b. Participatory Guarantee System for India (PGS-India)

iii. **Exemption-** Organic food which is marketed through direct sales to the end consumer by the small original producer or producer organisation is exempted from the provisions of the certification as mentioned above. Small original producer or producer organisation is the one whose annual turnover is not exceeding Rs 12 Lakhs per annum.

iv. **Labelling-** All organic foods shall comply with the packaging and labelling requirements specified under the Food Safety and Standards (Packaging and Labelling) Regulations, 2011 in addition to the labelling requirements under NPOP or PGS-India. Such products may carry a certification or quality assurance mark of NPOP or PGS-India in addition to the Food Safety and Standard Authority of India's organic logo (Jaivik Bharat Logo).

v. **Traceability-** Traceability to be established through one of the above mentioned systems.

vi. **Other relevant Regulations-** All Organic Food to comply with the requirements of product standards, as provided in the Food Safety and Standards (Food Product Standards and Food Additives) Regulations,

2011 except for additives which shall be as per NPOP or PGS-India. Such foods shall also comply with relevant provisions, as applicable under the Food Safety and Standards (Contaminants, Toxins and Residues) Regulations, 2011 except for residues of insecticides for which the maximum limits shall be 5% of the maximum limits prescribed or Level of Quantification (LoQ) whichever is higher.

vii. **Imports-**Organic food imports under bilateral or multilateral agreements on the basis of equivalence of standards between NPOP and the organic standards of the respective exporting countries shall not be required to be re-certified on import to India.

viii. **Retail-**The seller of organic food either exclusively or as part of his retail merchandise shall display such food in a manner distinguishable from the display of non-organic food.

Indian Organic Integrity Database Portal

Indian Organic Integrity Database Portal is a portal which has been developed by FSSAI. At the heart of Portal is the database of the all certified organic food operators under NPOP and PGS- India. Through this Portal the stakeholders/consumers can access all the information with respect to the certified operator, the certification system along with validity of certification, information on availability of certified organic products, other updates and resources on information pertaining to Organic Foods.

The Portal can be accessed at https://jaivikbharat.fssai.gov.in/

Chapter 12

Post Harvest Management of Organic Produces

Optimal quality organic produce that combines the desired textural properties, sensory shelf-life, and nutritional content results from the careful implementation of recommended production inputs and practices, careful handling at harvest, and appropriate postharvest handling and storage. This section will focus on an overview of general postharvest handling considerations unique to the marketing of registered or certified organic produce.

Maturity index for fruits and vegetables

The principles dictating at which stage of maturity a fruit or vegetable should be harvested are crucial to its subsequent storage and marketable life and quality. Post-harvest physiologists distinguish three stages in the life span of fruits and vegetables: maturation, ripening and senescence. Maturation is indicative of the fruit being ready for harvest. At this point, the edible part of the fruit or vegetable is fully developed in size, although it may not be ready for immediate consumption. Ripening follows or overlaps maturation, rendering the produce edible, as indicated by taste. Senescence is the last stage, characterized by natural degradation of the fruit or vegetable, as in loss of texture, flavour, etc. (senescence ends at the death of the tissue of the fruit). Some typical maturity indexes are described in following sections.

a) Skin colour

This factor is commonly applied to fruits, since skin colour changes as fruit ripens or matures. Some fruits exhibit no perceptible colour change during maturation, depending on the type of fruit or vegetable. Assessment of harvest maturity by skin colour depends on the judgment of the harvester, but colour charts are available for cultivars, such as apples, tomatoes, peaches, chilli peppers, etc.

b) Optical methods

Light transmission properties can be used to measure the degree of maturity of fruits. These methods are based on the chlorophyll content of the fruit, which is reduced during maturation. The fruit is exposed to a bright light, which is then switched off so that the fruit is in total darkness. Next, a sensor measures the amount of light emitted from the fruit, which is proportional to its chlorophyll content and thus its maturity.

c) Shape

The shape of fruit can change during maturation and can be used as a characteristic to determine harvest maturity. For instance, a banana becomes more rounded in cross-sections and less angular as it develops on the plant. Mangoes also change shape during maturation. As the mango matures on the tree the relationship between the shoulders of the fruit and the point at which the stalk is attached may change. The shoulders of immature mangoes slope away from the fruit stalk; however, on more mature mangoes the shoulders become level with the point of attachment, and with even more maturity the shoulders may be raised above this point.

d) Size

Changes in the size of a crop while growing are frequently used to determine the time of harvest. For example, partially mature cobs of *Zea mays saccharata* are marketed as sweet corn, while even less mature and thus smaller cobs are marketed as baby corn. For bananas, the width of individual fingers can be used to determine harvest maturity. Usually a finger is placed midway along the bunch and its maximum width is measured with callipers; this is referred to as the calliper grade.

e) Aroma

Most fruits synthesize volatile chemicals as they ripen. Such chemicals give fruit its characteristic odour and can be used to determine whether it is ripe or not. These doors may only be detectable by humans when a fruit is completely ripe, and therefore has limited use in commercial situations.

f) Fruit opening

Some fruits may develop toxic compounds during ripening, such as ackee tree fruit, which contains toxic levels of hypoglycin. The fruit splits when it is fully mature, revealing black seeds on yellow arils. At this stage, it has been shown to contain minimal amounts of hypoglycin or none at all. This creates a problem in marketing; because the fruit is so mature, it will have a very short post-harvest life. Analysis of hypoglycine 'A' (hyp.) in ackee tree fruit revealed that the seed contained appreciabl hyp. at all stages of maturity, at approximately 1000 ppm, while levels in the membrane mirrored those in the arils. This analysis supports earlier observations that unopened or partially opened ackee fruit should not be consumed, whereas fruit that opens naturally to over 15 mm of lobe separation poses little health hazard, provided the seed and membrane portions are removed.

g) Leaf changes

Leaf quality often determines when fruits and vegetables should be harvested. In root crops, the condition of the leaves can likewise indicate the condition of the crop below ground. For example, if potatoes are to be stored, then the optimum harvest time is soon after the leaves and stems have died. If harvested earlier, the skins will be less resistant to harvesting and handling damage and more prone to storage diseases.

h) Abscission

As part of the natural development of a fruit an abscission layer is formed in the pedicel. For example, in cantaloupe melons, harvesting before the abscission layer is fully developed results in inferior flavoured fruit, compared to those left on the vine for the full period.

i) Firmness

A fruit may change in texture during maturation, especially during ripening when it may become rapidly softer. Excessive loss of moisture may also affect the texture of crops. These textural changes are detected by touch, and the harvester may simply be able to gently squeeze the fruit and judge whether the crop can be harvested. Today sophisticated devices have been developed to measure texture in fruits and vegetables, for example, texture analyzers and pressure testers; they are currently available for fruits and vegetables in various forms. A force is applied to the surface of the fruit, allowing the probe of the penetrometer or texturometer to penetrate the fruit flesh, which then gives a reading on firmness. Hand held pressure testers could give variable results because the basis on which they are used to measure firmness is affected by the angle at which the force is applied.

j) Juice content

The juice content of many fruits increases as the fruit matures on the tree. To measure the juice content of a fruit, a representative sample of fruit is taken and then the juice extracted in a standard and specified manner. The juice volume is related to the original mass of juice, which is proportional to its maturity.

k) Sugars

In climacteric fruits, carbohydrates accumulate during maturation in the form of starch. As the fruit ripens, starch is broken down into sugar. In non-climacteric fruits, sugar tends to accumulate during maturation. A quick method to measure the amount of sugar present in fruits is with a brix hydrometer or a refractometer. A drop of fruit juice is placed in the sample holder of the refractometer and a reading taken; this is equivalent to the total amount of soluble solids or sugar content. This factor is used in many parts of the world to specify maturity. The soluble solids content of fruit is also determined by shining light on the fruit or vegetable and measuring the amount transmitted. This is a laboratory technique however and might not be suitable for village level production.

l) Starch content

Measurement of starch content is a reliable technique used to determine maturity in pear cultivars. The method involves cutting the fruit in two and dipping the cut pieces into a solution containing 4% potassium iodide and 1% iodine. The cut surfaces stain to a blue-black colour in places where starch is present. Starch converts into sugar as harvest time approaches. Harvest begins when the samples show that 65-70% of the cut surfaces have turned blue-black.

m) Acidity

In many fruits, the acidity changes during maturation and ripening, and in the case of citrus and other fruits, acidity reduces progressively as the fruit matures on the tree.

Planning for Postharvest Quality

Achieving an economically rewarding enterprise via the marketing of organic produce must begin well before harvest. Seed selection can be a critical decision factor in determining the postharvest performance of any commodity. Individual cultivars have variable inherent potential for firmness retention, uniformity, disease and pest resistance, and sensory shelf–life, to list a few key traits. Cultivars chosen for novelty or heirloom traits may be suitable for small-scale production and local marketing but would be disastrous if shipment to more distant markets was attempted. In addition to genetic traits, environmental factors such as soil type, temperature, wind during fruit set, frost, and rainy weather at harvest can have an adverse effect on storage life, suitability for shipping, and quality. Cultural practices may have dramatic impacts on postharvest quality. For example, poor seedbed preparation for carrots may result in sunburned shoulders and green core with many of the specialty carrots favoured by consumers at farmers markets.

Influence of Pre-harvest Factors on Postharvest Quality

Obtaining the optimum postharvest quality of vegetables actually begins very early in the farm planning process. The effects of preharvest factors on postharvest quality are often overlooked and underestimated. However, many of the decisions that we make during crop production can greatly influence the postharvest quality of crops. It is critical to remember that vegetable quality is only maintained postharvest – it is not improved during the harvest and storage processes. Thus, it is of utmost importance to consider the preharvest factors that allow us to maximize the quality of the vegetables going into storage. These factors encompass production and management decisions concerning soil fertility, variety selection, irrigation, and pest management.

a) Soil Factors

Maintaining good, long-term soil health and quality remains a primary goal of organic production systems. Achieving this goal will ultimately benefit the postharvest quality of vegetables grown on the farm, as the availability of the optimal levels of plant nutrients throughout the growing season will allow for optimal quality of the vegetables throughout the packing and distribution processes. Deficiencies or over abundances of certain plant nutrients can affect positively or negatively a crop's susceptibility to physiological disorders, disease, and negative composition and textural changes. When optimizing soil fertility to improve postharvest quality, it is important to remember that these may not be the same soil nutrient levels that produce the highest yield.

b) Nitrogen

Nitrogen is an important mineral element that is used by almost all crops. Nitrogen, as a key component of plant proteins, plays an important role in plant growth and development. Because of nitrogen's involvement in protein synthesis, soil nitrogen deficiencies may lead to lower protein concentrations in vegetables, thereby affecting the nutritional composition of the crop. Adequate soil nitrogen supplies allow for the optimal development of vegetable color, flavor, texture, and nutritional quality.

Excess soil nitrogen can be problematic as well. Research has shown that too much soil nitrogen can reduce the vitamin C content of green leafy vegetables. Excess nitrogen may lower fruit sugar content and acidity. In certain situations, leafy green plants may accumulate excess soil nitrogen, leading to high concentrations of nitrates in the harvested greens.

Specific examples of excess nitrogen negatively affecting crop quality include:

- Altered celery flavour
- "Brown-checking" of celery
- Weight loss of sweet potato during storage
- Hollow stem in broccoli
- Soft rot in stored tomatoes

c) Phosphorus and Potassium

Phosphorus and potassium also play very important roles in plant growth and development. Phosphorus is a key component of DNA and plant cell membranes. This element also plays a key role in plant metabolic processes. Potassium is important in plant water balance and enzyme activation. High levels of soil phosphorus have been shown to increase sugar concentrations of fruits and vegetables while decreasing acidity. High levels of soil potassium often have a positive effect on the quality of vegetables. Increased soil potassium concentrations have been shown to increase the vitamin C and titratable acidity concentrations of vegetables and improve vegetable colour. Potassium also decreases blotchy ripening of tomato.

d) Calcium

Elemental calcium is an important to plant cell walls and membranes. Deficiencies in soil calcium have been associated with a number of postharvest disorders, including blossom end rot of tomato, pepper, and watermelon; brown-heart of escarole; blackheart in celery; and tip-burn of lettuce, cauliflower and cabbage. High soil calcium concentrations reduce these disorders and are associated with other postharvest benefits, including increased vitamin C content, extended storage life, delayed ripening, increased firmness, and reduced respiration and ethylene production

e) Soil Texture

The texture of the soil on which certain vegetable crops are grown may also affect the post harvest quality. For example, carrots grown on muck soils has shown a greater concentration of a chemical that imparts a bitter flavour, than carrots grown on sandy soil.

f) Irrigation

Adequate soil moisture during the pre-harvest period is essential for the maintenance of post harvest quality. Water stress during the growing season can affect the size of the harvested plant organ, and lead to soft or dehydrated fruit that is more prone to damage and decay during storage. On the other hand, vegetables experiencing an excess of water during the growing season can show a dilution of soluble solids and acids, affecting flavor and nutritional quality.

Excess moisture on the harvested vegetable can also increase the incidence of post harvest diseases. To minimize the amount of water on the harvested vegetable brought into storage, it may be beneficial to choose surface or subsurface irrigation rather than overhead irrigation. Vegetables harvested in the early morning, during rainy periods, and from poorly ventilated areas can also experience increased post harvest decay.

g) Insect Pests

Insect pest problems during the growing season can also affect postharvest quality, both obvious and no-so obvious ways. Visible blemishes on the vegetable surface caused by insect feeding can have a negative effect on the appearance of vegetables, thus decreasing their appeal to consumers. Feeding injury on vegetables by insects can lead to surface injury and punctures, creating entry points for decay organisms and increasing the probability of postharvest diseases. In addition, the presence of insect pests on vegetables entering storage leads to the possibility of these insects proliferating in storage and becoming an issue.

h) Selection of Vegetable Varieties

The selection of the right vegetable variety for your farm and market channel can greatly influence the subsequent postharvest quality. Certain varieties are more suited for the longer-term storage that is essential for marketing to larger wholesale outlets. Other varieties may optimize taste, essential for the post-harvest quality of vegetables going to farmers markets.

When planning which vegetable varieties to grow on your farm, it is important to consider which harvest windows are needed. Vegetables harvested at the incorrect stage of maturity will have a significant decrease in postharvest quality. Quality characteristics such as texture, fibre and consistency are greatly affected by stage of maturity at harvest. Fruits and vegetables that are harvested while immature are highly susceptible to shrivelling and mechanical damage. Fruits and vegetables harvested at an overripe stage often have poor texture and flavour. Suboptimal harvest dates lead to a greater susceptibility to post-harvest physiological disorders than harvest dates closer to the proper stage of maturity.

i) Other Production Considerations

Certain production techniques can also help to attain the optimal postharvest quality. These techniques include:

- Staking of tomato crops allows the fruit to remain off the ground during the growing season. By keeping the fruit off the ground, fruit blemishes and decay are minimized. This, in turn, leads to less postharvest decay in storage. Certain staking techniques also may allow more light penetration and air circulation through the canopy, increasing fruit yield and size.
- Pruning certain crops (such as tomatoes) can alter the microclimate around the plants in ways that benefit postharvest quality. For instance, removing some of the plant foliage can allow for better air circulation and thus minimize excess moisture around the fruits, leading to less decay and postharvest disease issues.

- Row covers over leafy greens can minimize physical damage to certain vegetables, especially leafy greens. By minimizing physical injury to the plant tissue, fewer entry points for microorganisms are present on the vegetable surface, thus minimizing the potential for post harvest diseases to manifest.

Harvest Handling

The inherent quality of produce cannot be improved after harvest, only maintained for the expected window of time characteristic of the commodity. Part of successful postharvest handling knows what this window of opportunity is under your specific conditions of production, season, method of handling, and distance to market. Among the benefits of organic production, it is often more common to harvest and market near or at peak ripeness than in many conventional systems. However, organic production often includes more specialty varieties that have reduced or even inherently poor shelf life and shipping traits. As a general approach, the following practices can help to maintain quality:

- Harvest during the coolest time of the day to maintain low product respiration.
- Avoid unnecessary wounding, bruising, crushing, or damage from humans, equipment, or harvest containers.
- Shade harvested product in the field to keep it cool. Covering harvest bins or totes with a reflective pad greatly reduces heat gain from the sun and reduces water loss and premature senescence.
- If possible, move product into a cold storage facility or postharvest cooling treatment as soon as possible. For some commodities, such as berries, tender greens and leafy herbs, one hour in the sun is too long.
- Don't compromise high quality product by intermingling damaged, decayed, or decay-prone product in a bulk or packed unit.
- Only use cleaned and, as necessary, sanitized packing or transport containers.

These operating principles are important in all operations but carry special importance for many organic producers due to limited post harvest cooling opportunities.

Table 18. Maturity indices for fruits

Index	Examples
Elapsed days from full bloom to harvest	Apples, pears
Mean heat units during development	Peas, apples, sweet corn
Development of abscission layer	Some melons, apples, feijoas
Surface morphology and structure	Cuticle formation on grapes, tomatoes Netting of some melons Gloss of some fruits (development of wax)
Size	All fruits and many vegetables
Specific gravity	Cherries, watermelons, potatoes
Shape	Angularity of banana fingers Full cheeks of mangos Compactness of broccoli and cauliflower
Solidity	Lettuce, cabbage, Brussels sprouts
Textural properties	
Firmness	Apples, pears, stone fruits
Tenderness	Peas
Color, external	All fruits and most vegetables
Internal color and structure	Formation of jelly-like material in tomato fruits Flesh color of some fruits
Compositional factors	
Starch content	Apples, pears
Sugar content	Apples, pears, stone fruits, grapes
Acid content, sugar/acid ratio	Pomegranates, citrus, papaya, melons, kiwifruit
Juice content	Citrus fruits
Oil content	Avocados
Astringency (tannin content)	Dates
Internal ethylene concentration	Apples, pears

Table 19. Maturity indices for vegetables and root crops.

Crop	Index
Root, bulb and tuber crops	
Radish and carrot	Large enough and crispy (over-mature if pithy)
Potato, onion, and garlic	Tops beginning to dry out and topple down
Yam, bean and ginger	Large enough (over-mature if tough and fibrous)
Green onion	Leaves at their broadest and longest
Fruit vegetables	
Cowpea, yard-long bean, snap bean, sweet pea, and winged bean	Well-filled pods that snap readily
Lima bean and pigeon pea	Well-filled pods that are beginning to lose theirgreenness
Okra	Desirable size reached and the tips of which can be snapped readily
Snake gourd, and dishrag gourd	Desirable size reached and thumbnail can still penetrate flesh readily (over-mature if thumbnail cannot penetrate flesh readily)
Eggplant, bitter gourd, cucumber	Desirable size reached but still tender (over mature if color dulls or changes and seeds are tough)
Sweet corn	Exudes milky sap when thumbnail penetrates kernel
Tomato	Seeds slipping when fruit is cut, or green color turning pink
Sweet pepper	Deep green color turning dull or red
Muskmelon	Easily separated from vine with a slight twist leaving clean cavity
Honeydew melon	Change in fruit color from a slight greenish white to cream; aroma noticeable
Watermelon	Color of lower part turning creamy yellow, dull hollowsound when thumped
Flower vegetables	
Cauliflower	Curd compact (over mature if flower cluster elongates and become loose)

Broccoli	Bud cluster compact (over mature if loose)
Leafy vegetables	
Lettuce	Big enough before flowering
Cabbage	Head compact (over mature head cracks)
Celery	Big enough before it becomes pithy

Harvesting time

When the decision to harvest has been taken the preferred time for harvest varies from crop to crop. However the preferred time is the coolest part of the day, usually in the early morning or late afternoon. This is particularly so for leafy vegetables.

Other factors such as the availability of labor and transport and the distance to a packing house or temporary storage area may dictate that some other harvesting time is more suitable or necessary.

The selected time should be that which minimize the time between harvest and transport to a packing shade. For example, if night transportation is used it is not advisable to harvest early morning unless produce can be placed under cover and a well ventilated place during the daytime.

Local weather conditions could affect the harvesting time:

- It is not desirable to harvest produce when it is wet from dew or rain as this greatly increases the risk of post-harvest spoilage and the tissue is more prone to physical damage.
- It is also not advisable to harvest during a hot and sunny day if the produce is left in the field and cannot be protected by the sunrays and the heat.
- Protect harvested produce in the field by putting it under open-sided shade when transport is not immediately available. Produce left exposed to tropical sunlight will get very hot. Leafy green vegetables such as spinach and salad lose water quickly because they have a thin waxy skin with many pores.

Harvesting technique and operations

In developing countries most of the produce for internal rural and urban markets is harvested by hand. Larger commercial producers may find a degree of mechanization an advantage, but the use of sophisticated harvesting machinery will be limited for the most part to agro-industrial production of cash crop for processing or export. In most circumstances, if harvesting properly is done by hand by trained and experienced workers, will result in less damage than if produce is machine harvested. Hand-harvesting is usually preferred when fruits, such as peaches, and other produce, such as green vegetables, are at different stages of maturity and there is need on repeated visits to harvest the crop over a period of time.

Fruits

Many ripe fruits such as apples and. some immature seed-bearing structures such as legumes pods have a natural breakpoint of the fruit stalk, which can be easily broken by twisting and lifting the stalk taken between the thumb and index. Fruits and other seed-bearing structure harvested in the immature or unripe green stage are more difficult to pick without causing damage

Plucking methods vary according to the kind of produce being harvested:

- Ripe fruits with a natural break – point, which leave the stalk attached to the fruit are best removed by a "lift, twist and pull" series of movements, e.g. tomato and passion fruit.
- Mature green or ripe fruits with woody stalks which break at the junction of the fruit and the stalk are best clipped from the tree, leaving up to a centimeter of fruit stalk attached. If the stem is broken off at the fruit itself, diseases may enter the stem scar and give rise to stem end rot, e .g. mango, citrus, avocado;
- Immature fruits with fleshy stem can be cut with a sharp and clean knife, e.g., okra, papaya, capsicum; these can also be harvested by breaking the stem by hand, with the risk of damaging the plant or the fruit and the rough cut will make the produce and the plant more susceptible of decay.

Harvesting is basically a simple operation involving the removal of the produce from the parent plant and placing it in containers for removal from the field to the market, or to the packing shade in the farm itself or to the packing – house.

Vegetables

Either the whole or a part of vegetative growth can be harvested byhand only or sharp knife. Knives may be kept sharp and clean at all times to avoid spreading of virus diseases from plant to plant.

Harvesting methods varies in accordance with the plant part harvested:

- Leaves only (spinach, rape, etc.) and lateral buds (Brussels sprouts, etc): the stem is snapped off by hand;
- Above–ground part of the plant (cabbage, lettuce, etc.): the main stem is cut through with a heavy knife, and trimming (roots and external unsuitable leafs are discarded) is done in the field. Do not forget that the cut stem must not be placed in the soil;
- Immature green onions can usually be pulled from the soil by hand; leek, garlic and mature bulb onions are loosened by using a digging fork as for root crops such as carrots and lifted by hand. Simple tractor implements are available for undermining bulbs and bringing them to the surface.

Flower structure vegetables

Immature flower heads (cauliflowers, broccoli) can be cut with a sharp knife and trimmed in the field. Mature flowers (squash, chayote, pumpkins) are plucked individually by hand or shoot bearing flowers are harvested as a vegetable.

Root and tuber crops

Most roots and tubers that live beneath the soil are likely to suffer mechanical damage ay harvest because of digging tools, which may be wooden sticks, machetes, hoes or forks. Harvesting of those crops is easier if they are grown on raised beds or mounds, or “earthed up” as is common with potato growing. This enables the digging tool to be pushed into the soil under the roots or tubers, which then can be levered upward, loosening the soil and decreasing the possibility of damage to the crop.

Other root crops, such as taro, carrots, turnips, radishes, etc. can be loosened from the soil at an angle and leaving the root upward. This method can be used also for celery if it has been earthed up or buried to blanch the stem.

Post-harvest transport

Transport of the produce can be affected as follows:

Field and farm transport

Routes for the movement of produce within farm fields should be planned before crops are planted. Farm roads should be kept in good conditions because great damage can be inflicted on produce carried over rough roads in unsuitable vehicles. Containers must be loaded in the transport vehicle carefully and stacked in such a way that they cannot shift or collapse, damaging the content. Vehicles need good shock absorbers and low-pressure tires and must move with care.

Transport from the farm

The destination of the produce leaving the farm will usually be one of the following:

- A local market. Produce is usually in small containers carried sometimes by animals or in animal drawn carts; public transport is sometime used .Usually produce is graded and packed in the field.
- A commercial packing house. Produce may be in palletized field containers or in hand loaded sacks or wooden or plastic boxes. In the packing-house the produce is graded and packed in suitable containers for the market.
- A city market. This applies only where produce is graded and packed in marketing containers on the farm or the packinghouse.

Postharvest handling

Temperature is the single most important tool to maintain postharvest quality. Other than field cured or durable products, removing field heat as rapidly as possible is highly desirable. Harvesting cuts off a vegetable from its source of water. However, it is still alive and will lose water, and therefore turgor, due to respiration. Field heat can accelerate the rate of respiration and therefore the rate of quality loss. Proper cooling protects quality and extends both the sensory (taste) and nutritional shelf life of produce. The capacity to cool and store produce creates greater market flexibility. There is a tendency by growers to underestimate the refrigeration capacity needed for peak cooling demand. It is often critical to reach the desired short-term storage or shipping pulp temperature rapidly to maintain the highest visual quality, flavor, texture, and nutritional content of fresh produce. The most common cooling methods are:

1. **Room cooling**: An insulated room or mobile container equipped with refrigeration units. Room cooling is slow compared with other options. Depending on the commodity, packing unit, and stacking arrangement the product may cool too slowly to prevent water loss, premature ripening, or decay.
2. **Forced-air cooling**: Fans are used in conjunction with a cooling room to pull cool air through packages of produce. Although the cooling rate depends on the air temperature and the rate of airflow, this method is usually 75–90% faster than room cooling.
3. **Hydro-cooling**: Showering produce with chilled water is an efficient way to remove heat, and can serve as a means of cleaning at the same time. Use of a disinfectant in the water is essential and the some of the currently permitted products are discussed in the following section. Hydro-cooling is not appropriate for all produces. Water proof containers or resistant waxed-corrugated cartons are required. There are also problems with the requirement for a large quantity of clean water, disposing of waste water, heavy capital cost and it is not applicable to all types of packaging especially cartons. This technique is used for small fruits or vegetables, leafy vegetables

and pineapple. Currently waxed corrugated cartons have limited recycling or secondary use outlets and reusable, collapsible plastic containers are gaining popularity.

4. **Top or liquid icing**: Icing is an effective method to cool tolerant commodities and is equally adaptable to small or large-scale operations. Ensuring that the ice is free of chemical, physical, and biological hazards is essential.
5. **Vacuum cooling**: Under vacuum, water within the plant evaporates and removes heat from the tissues. This system works well for leafy crops, such as lettuce, spinach, and celery, which have a high surface-to-volume ratio. Water may be sprayed on the produce prior to placing it vacuum. As with hydro-cooling, proper water disinfection is essential. The cost of the vacuum chamber system restricts its use to larger operations.

Packing and packaging materials

The IFOAM Draft Basic Standards 2002 state that 'Organic product packaging should have minimal adverse environmental impacts'; and recommend that 'Processors of organic food should avoid unnecessary packaging materials; and organic food should be packaged in reusable, recycled, recyclable and biodegradable packaging whenever possible'.

Thus, although all types of packaging are authorized, there is an expectation that careful thought will have gone into the choice of the packaging with regard to its environmental impact. In the future, restrictions may be put in place concerning the use of packaging materials that are harmful to the environment, especially for those packaging materials that are not recyclable or biodegradable.

Natural material: Baskets and other traditional containers are made from raw material locally available such as bamboo, rattan, straw and palm leaves. Both raw material and labor costs for the manufacture of the containers are normally low, and if the containers are well made, they can be reused for several times. They are mostly used for handling the produce in the field, at farm level and seldom for the local market.

Disadvantages are:

- They are difficult to clean and easily contaminated;
- They lack rigidity and easily bend out of shape when stacked;
- They load badly for their shape;
- They cause pressure damage if tightly filled; and
- They often have sharp edges or splinters and cause damage unless they internally lined with sacks, cotton or other materials.

Wood: Sawn wood is used to manufacture reusable boxes or crates,but less so recently because of cost. Veneers of various thicknesses are used to make various boxes or trays. Wooden boxes are rigid and reusable and if made to a standard size, it will stack well on trucks and in storage. They are used for produce handling at farm and packing – house levels.

Disadvantages are:

- Difficult to clean adequately for multiple uses;
- Heavy and costly to transport for the reutilization; and
- They have sharp edges, requiring some form of liner to protect the content.

Cardboard and paper

Whilst these traditional materials are generally readily available and inexpensive, they have several drawbacks: porous to gas, permeable to water, easily torn or crushed. They protect products only from light impacts. In organic farming, these materials are principally used for fresh fruits and vegetables. In order to limit impacts between products and to limit movement within the packaging, the use of liners between layers of fruits and vegetables, or of individual paper wrapping, can be efficient. Waxing of packaging restricts water permeability but can make the package unsuitable for recycling. Particular consideration needs to be paid to the use for which the packaging is intended. A lightweight cardboard box may be adequate for use in the local market, but a telescopic box with reinforced corners may be necessary for sea freight. It will need to retain its strength during extended periods at low temperatures and high humidity's to enable stacking over 2 m high on a pallet.

Plastic

Plastic packaging is likely to deliver the best quality produce, minimising wastage. It can be pre-printed for marketing purposes and it is ideally suited to forming flexible, unbreakable packaging matched to the product's needs. It is light in weight, leading to cheaper transport costs and less fuel consumption in transport. Although plastic packaging is often frowned upon because it is commonly derived from fossil fuels and is not always able to be recycled, the alternatives need to be considered carefully. There are many recyclable or bio-degradable plastics and some plastics are now produced from starch.

Plastic is selectively permeable to gas and water, depending on the type of polymer. Some polymers are therefore ideally suited to creating a modified atmosphere around fresh fruits and vegetables. During storage, the respiration of the produce emits carbon dioxide and consumes oxygen. The selective permeability of the polymer results in an increase of carbon dioxide inside the packaging. The respiration rate of the vegetable decreases proportionally to this enrichment, and the composition of the atmosphere stabilize progressively. The resultant atmosphere, if maintained, can lead to significant improvements in storage quality.

This type of modified atmosphere packaging (MAP) is being developed for highly perishable, high-value produce. The problem is that temperature fluctuations dramatically affect the rates of tissue metabolism and the permeability of the plastics; stable gas compositions can be obtained only in a precisely controlled environment. The risk of anaerobism in MAP is severe; anaerobism leads to the production of off-flavours and, in extreme cases, favours the growth of toxic organisms and toxin production. MAP is thus too risky to recommend for routine use in the export of fresh fruit and vegetables from developing countries.

Glass

Glass receptacles are principally used for liquid products or solids in liquid. Glass receptacles are well adapted to organic products as they are impermeable to gas, air moisture, micro-organisms and resistant to thermal treatments. Against these positive attributes are the bulk and transparency of glass, which can cause problems for products that are sensitive to light. Also, the energy costs of recycling glass are very high (possibly making it less ecologically-friendly than plastic). In addition, glass breakage is extremely serious on packing lines; a single breakage can lead to expensive downtime as equipment is turned off and cleaned to prevent glass shards entering packages.

Metal

Metal offers the same sealing advantages and resistance to thermal treatments as glass. However, the consumer image of this type of packaging is unfavourable; and metal can be subject to corrosion.

Among the four types of packaging, the most appropriate for organic produce storage is often plastic or glass. In practice, combinations of containers are often used, e.g. glass containers in cardboard cartons. For processing in developing countries, packaging materials often have to be imported from industrialised countries. This implies constraints and supplemental costs (management of stock, financial investments) that can hinder the development and marketing of organic products.

Storage

The term "storage" as now applied to fresh produce is almost automatically assumed to mean the holding of fresh fruit and vegetables under controlled conditions. Although this includes the large-scale storage of some major crops, such as carrots and potatoes, to meet a regular continuous demand and provide a degree of price stabilization it also meets the demands of populations of developed countries and of the richer consumers of developing countries, providing year round availability of various local and exotic fruits and vegetables of acceptable quality.

In many developing countries, however, where seasonally produced plant foods are held back from sale and released gradually, storage in a controlled environment is not possible because of the cost and the lack of infrastructure development and maintenance and managerial skill. Even in developed countries, however, there are still many peoples who, for their own consumption, preserve and store fresh produce by traditional methods.

Storage potential

Much fresh produce (i.e. that which is most perishable) cannot be stored without refrigeration, but the possibilities of extending the storage life of even the most durable fresh produce under ambient conditions are limited.

Organs of survival

The organs of survival which forms the edible parts of many crops such as potatoes, yams, beets, carrots and onions have a definite period of dormancy after harvest and before they resume growth, at which time their food value declines. This period of dormancy can usually be extended to give the longest possible storage if appropriate conditions are provided. This factor is called the storage potential. It is important to recognize the variation in the storage potential of different cultivars of the same crop. Experienced local growers and seed suppliers can usually provide information on this subject.

Edible reproductive parts

These is largely confined to the fruits or seeds of leguminous plants (peas and beans). In their fresh condition these products have brief storage life which can be only slightly extended by refrigeration. They can also be dried and then are called pulses. Pulses have a long storage life, provided they are kept dry, and do not present a storage problem as is the case of fresh produce.

Fresh fruits and vegetables

These include the leafy green vegetables, fleshy fruits and modified flower plants.(e.g. cauliflower and pineapple). The storage potential of these, particularly tropical fruits in tropical countries, is very limited under ambient conditions. They quickly deteriorate because of their fast respiration rates, which cause rapid heat build-up and depletion of their high moisture content.

Most fresh fruits and vegetables have a storage life of only a few days under even the best environmental conditions.

Recommended storage temperatures

Table 20. Recommended temperature and relevant humidity and storage life for fruits, vegetables and root crops.

Product	Temperature		Relative humidity (%)	Approximate storage life
	°C	°F		
Apples	1-4	30-40	90-95	1-12 months
Apricots	-0.5-0	31-32	90-95	-1-3 weeks
Asian Pear	1	34	90-95	5-6 months
Avocados, Fuerte, Hass	7	45	85-90	2 weeks
Avocados, Lula, Booth-1	4	40	90-95	4-8 weeks
Avocados, Fuchs, Pollock	13	55	85-90	2 weeks
Bananas, green	13-14	56-58	90-95	14 weeks
Barbados Cherry	0	32	85-90	7-8 weeks
Bean sprouts	0	32	95-100	7-9 days
Beans dry	4-10	40-50	40-50	6-10 months
Beans, green or snap	4-7	40-45	95	7-10 days
Beans, Lima, in pods	5-6	41-43	95	5 days
Beans, bunched	0	32	98-100	10-14 days
Beets, topped	0	32	98-100	4-6 months
Bitter melon	12-13	53-55	85-90	2-3 weeks
Black sapota	13-15	55-60	85-90	2-3 weeks
Blood Orange	4-7	40-44	90-95	3-8 weeks
Broccoli	13-15	55-60	85-90	2-6 weeks
Cabbage, early	0	32	95-100	10-14 days
Cabbage, late	0	32	98-100	3-6 weeks
Carambola	9-10	48-50	58-90	3-4weeks
Carrots, bunched	0	32	95-100	2 weeks
Carrots, mature	0	32	98-100	7-9 months
Carrots, immature	0	32	98-100	4-6 weeks
Cashew apple	0-2	32-36	85-90	5 weeks
Cauliflower	0	32	95-98	34 weeks
Celery	0	32	98-100	2-3 months
Cherries, sour	0	32	90-95	3-7 days
Cherries, sweet	-1 to 0.5	30-31	90-95	2-3 weeks
Coconuts	0-1.5	32-35	80-85	1-2 months
Corn, sweet	0	32	95-98	5-8 days
Cucumbers	10-13	50-55	95	10-14 days
Custard apples	5-7	41-45	85-90	46 weeks
Eggplant	12	54	90-95	1 week

(Contd.)

Endive and escarole	0	32	95-100	2-3 weeks
Garlic	0	32	65-70	6-7 months
Ginger root	13	55	65	6 months
Granadilla	10	50	85-90	3-4 weeks
Grapefruit, Calif. & Ariz	14-15	58-60	80-90	6-8 weeks
Grapefruit, Fla.& Texas	10-15	50-60	80-90	68 weeks
Greens, leafy	0	32	95-100	10-14 days
Guavas	5-10	41-50	90	2-3 weeks
Jackfruit	13	55	85-90	2-6 weeks
Kiwifruit	0	32	90-95	3-5 months
Leeks	0	32	95-100	2-3 months
Lettuce	0	32	98-100	2-3 weeks

Table 21: Recommended temperature and relevant humidity and storage life for fruits, vegetables and root crops.

Product	Temperature		Relative humidity (%)	Approximate storage life
	°C	°F		
Lime	9-10	48-50	85-90	6-8 weeks
Malanga	7	45	70-80	3 months
Mangoes	13	55	85-90	2-4 weeks
Crenshaw	7	45	90-95	2 weeks
Honeydew	7	45	90-95	3 weeks
Persian	7	45	90-95	2 weeks
Okra	7-10	45-50	90-95	7-10 weeks
Onion, green	0	32	95-100	34 weeks
Onion, dry	0	32	65-70	1-8 months
Onion set	0	32	65-70	6-8 months
Oranges, Calif. & Ariz	3-9	38-48	85-90	3-8 weeks
Oranges, Fla. & Texas	0-1	32-34	85-90	8-12 weeks
Papayas	7-13	45-55	85-90	1-3 weeks
Passion fruit	7-10	45-90	85-90	3-5 weeks
Pears	-1.5 to 0.5	29-31	90-95	1-2 weeks
Peas, green	0	32	95-98	6-8 weeks
Peppers, Chill (dry)	0-10	32-50	60-70	6 months
Peppers, Sweet	7-13	45-55	90-95	2-3 weeks
Pineapples	7-13	45-55	85-90	24 weeks

(Contd.)

Plantain	13-14	55-58	90-95	1-5 weeks
Pummelo	7-9	45-48	85-90	12 weeks
Pumpkins	10-13	50-55	50-70	2-3 months
Sapodilla	16-20	60-68	85-90	2-3 weeks
Snow peas	0-1	32-34	90-95	1-2 weeks
Soursop	13	55	82-90	1-2 weeks
Spinach	0	32	95-100	10-14 days
Squashes, Summer	5-10	41-50	95	1-2 weeks
Squashes, winter	10	50	50-70	2-3 months
Sugar apples	7	45	85-90	4 weeks
Sweet Potatoes	13-15	55-60	85-90	4-7 months
Tamarinds	7	45	90-95	3-4 weeks
Tangerines, mandarins, & related	4	40	90-95	24 weeks
Taro root	7-10	45-50	85-90	4-5 months
Tomatoes, mature- green	18-22	65-72	90-95	1-3 weeks
Tomatoes, firm-ripe	13-15	55-60	90-95	4-7 days
Watermelons	10-15	50-60	90	2-3 weeks
White Sapot	19-21	67-70	85-90	2-3 weeks
Yams	16	61	70-80	6-7 months

Marketing

Organic farmers who produce for the market are interested to know the market potential and how to get access to organic markets. Giving best value to high quality organic products is a major concern of organic farmers and needs specific techniques. Marketing products as organic also requires certification of the farm. No organic certification is required, if the farm products are not sold as organic. The decision to certify the farm as organic should be linked to the possibility of marketing a relevant share of the farm products as organic with a premium price. The premium price should cover the certification costs.

Application for certification can be made, when the entire farm is managed organically. Depending on the organic standards there is a defined transition or conversion period of one to three years. During this time, depending on the standards, the farm products must either be marketed as non-organic, or they can be marketed as organic products originating from a farm in conversion. Most customers in export markets, however, request organic products that originate from farms that have already achieved the conversion period.

Marketing organic products involves considerable personal initiative. To access domestic and local markets, farmers need to communicate the value of their products to local traders and customers. This may involve inviting them to the farm and explaining the principles of organic production and showing them advantages of the organic approach for nature and the positive impact on product quality. Traders and customers buy organic products based on a certificate from an organic certification body.

What is value addition?

Value addition is a process in which for the same volume of a primary product, a high price is realized by means of processing, packing, upgrading the quality or other such methods.

What is value added agriculture?

Value-added agriculture refers most generally to manufacturing process that increases the value of primary agricultural commodities. Value-added agriculture may also refer to increasing the economic value of a commodity through particular production process, eg., organic produce, or through regionally branded products that increase consumer appeal and willingness to pay a premium over similar but differentiated products. Value-added agriculture is regarded by some, a significant rural development strategy. Small scale processing unit, organic food processing, non-traditional crop production, agri-tourism and bio-fuels development are examples of various value-added projects that have created new jobs in some rural areas.

Need for value addition

1. To improve the profitability of farmers
2. To empower the farmers and other weaker sections of society especially women through gainful employment opportunities and revitalize rural communities.
3. To provide better quality, safe and branded foods to the consumers.
4. To emphasize primary and secondary processing.

5. To reduce post harvest losses.
6. Reduction of import and meeting export demands.
7. Way of increased foreign exchange.
8. Encourage growth of subsidiary industries.
9. Reduce the economic risk of marketing.
10. Increase opportunities for smaller farms and companies through the development of markets.
11. Diversify the economic base of rural communities.
12. Overall, increase farmers' financial stability.

Market forces for product differentiation and value addition

1. Increased consumer demands regarding health, nutrition and convenience.
2. Efforts by food processors to improve their productivity.
3. Technological advances that enable producers to produce what consumers and processors desire.

Producers have a challenge to be responsive to consumer demands by producing what is desired. Attentiveness to consumer demands in quality, variety and packaging are important because demographic trends show growth in the convenience-oriented, health conscious and environmentally concerned sectors where price is not as important as quality.

Horticulture as a mean for value addition

1. Horticulture deals a large group of crops. Therefore, cultivation of crops which belong to us and possess great medicinal, nutritional, health promoting values.
2. India as second largest producer of fruits and vegetables, only 10 per cent of that horticultural produce is processed, but other developed and developing countries where 40-80 per cent produce is value added.
3. Horticultural crops provide varied type of components, which can be effectively and gainfully utilized for value addition like pigment, amino acids, oleoresins, antioxidants, flavors, aroma etc.

4. Post harvest losses in horticultural produce are 5 to 30 per cent which amounts to more than 8000 crore rupees per annum. If we subject our produce to value addition the losses can be checked.
5. Horticultural crops are right material for value addition because they are more profitable, has high degree of process ability and richness in health promoting compounds and higher potential for export.

Therefore, horticultural crops are right material for value addition in the present context of agricultural scenario and we should go for new product development to be unique and novel.

Value addition as new product development

To be unique and novel new product development should be attempted and this can be approached through various ways.

1. A product entirely new in character.
2. A product apparently similar in character in many respects to some existing brands but with distinguishing features.
3. A product similar to one already in the market but new form of manufacture
4. A product novel in kind, made by novel process or through novel ingredient.
5. A product resulting from substantial modifications of the characters with change in nature/proportion of ingredients or processing methods or conditions or packing system.

Some of the areas of achievement

1. Carnation -Enhanced shelf-life
2. Carnation - Modified flower color, sulphonyl urea, herbicide tolerance
3. Melons - Delayed ripening
4. Papaya - Resistance to viral infection, papaya ring spot virus (PRSV)

5. Squash - Resistance to watermelon mosaic virus and cucumber mosaic virus and herbicide tolerance to glufosinate ammonium.
6. Sugar beet - Herbicide tolerance
7. Tomato - Delayed ripening, resistance to lepidopteran pests, delayed softening, lycopene rich tomato.

Choice for value addition

Any attempt for value addition should focus for following parameter for deriving maximum benefit.

1. **Unique:** The product we develop should be one of its own kinds for which crop and variability indigenous to our country should be exploited.
2. **Novelty**: The product should be new and unusual like blue or black rose and likewise so that no one can compete.
3. **Export potential**: The product developed should have demand in international market for higher return and appreciation of benefit of global trade.
4. **High value**: The product should have high value for low volume for ease of trading and distribution and the extracts from Indian spices and herbal medicinal plants can fulfill this requirement.
5. **Availability**: Consistent availability of the product in required quantity should be ensured for stable market and faith.
6. **Market**: Any product that is developed must have market because market is the key for success of any product.

Chapter 13

Problem Soil Reclamation

Regeneration capacity of the soil is enhanced with the sowing of MVS which provides ideal condition for regeneration. It creates green cover / dry leaf cover in the top soil (35-40 cm height) which will facilitate the multiplication of inoculated beneficial microbes in the soil.

Multi Varietal seeds sowing techniques

MVS contains seeds of three to four crops belonging to each category viz., cereals, pulses oilseeds, nitrogen fixing green manures and spices and condiments. The seeds of above crops may be mixed in different proportion based on their growth habit and seasonal requirement. For example in the summer the following mixture can be practiced with each of 1-1.5 kg each.

Crop mixture for sowing

Cereals	Sorghum, Pearlmilet; Foxtailmillet, Kodomillet
Pulses	Blackgram, Greengram, Cowpea, Redgram
Oilseeds	Soybean, Castor, Sesame, Sunflower
N giving GM	Sunnhemp, Dhaincha, Wild indigo, *Sesbania* spp.
Spices and condiments	Chillies, Coriander and Aniseed

Take up sowing using appropriate crop mixtures at the rate of 20-25 kg/acre.

Under normal soil conditions

At the time of MVS sowing, 100 kg of well decomposed cow dung manure + palm sugar (1-3 kg) slurry spray to create more friable soil texture which conducive for the microbes. The above preparation may be split into equal four parts and inoculate each one with the bio-agents separately and allow it for 30 days to multiply the microbes.

After that, take a part of digested manure in to a gunny bag and place it in the entry point of irrigation water for slow dissolution and uniform spread into the field. Another part may be applied as basal dressing to soil at an interval of 30 days from the previous application. All manures get converted into available nutrients within 60 days after the incorporation of MVS.

Under inert / dead soil conditions

Due to the intensive cultivation using modern varieties with indiscriminate use of chemicals heavily depleted the organic matter content of the soil. Low organic matter content may not support the growth of microbes. Soil which having poor physical conditions and biological activity with low buffering capacity may be called a sick or dead soil. These soils can be revived for cultivation within a year through MVS technique as detailed below.

- Apply FYM / Poultry manure as basal
- First MVS sowing
- Allow it to grow for 30 DAS
- Inoculate microbial cultures and incorporate on 30 DAS
- Second MVS sowing, inoculate microbial cultures and incorporate on 60 DAS
- Third MVS sowing and keep it for 100-110 DAS, collect seeds and incorporate the entire biomass.

Methodology

- Applying bio-digested gas slurry (BDGS) (25 litres) + Cow's urine (5 litres) + 10 lit of water as slurry mixture applied @ 1050 litres / acre along with *irrigation water*.

- Bio-digested gas slurry (BDGS) + RK (Decayed fruits fermented in BDGS for a week and applied as *foliar spray* @ 10%)

Soil management

- For soil improvement in 'theri' lands of Tuticorin district, 200 tonnes to tank silt are applied per acre followed by 50 tonnes per year for the next few years (Farmers of Erode and Tuticorin District in Tamil Nadu)
- Crop residues and tree branches are burnt on the soil surface at the end of summer to improve the structure of clayey soil and to make ploughing easy (Farmers of Konkan Region in Maharashtra)

Different methods for soil conservation

Coir pith waste, farm waste, dried leaves, dried grasses, sugarcane trashes and groundnut shell can be used as soil conserving agents.

After the harvest of sugarcane crop, the trashes should be spread over on the ridges for conserving the moisture.

After planting banana, well decomposed sugarcane trashes can be spread over.

Problem soils

Physical problems

An optimum physical environment of soils is essential for better growth of plants, consequently for better yields. Based on soils physical properties viz., infiltration, bulk density, hydraulic conductivity, porosity (capillary and non capillary, aggregates etc soil physical constraints are identified as below.

1. Slow permeable soils
2. Excessively permeable soils
3. Subsoil hardening
4. Surface crusting

5. Fluffy paddy soils
6. Shallow soils

Slow permeable soils

Slow preamble soils are those having infiltration rates less than 6 cm/day due to high clay content of the soil. Due to low infiltration rates, the amount of water entering the soil profile is reduced thus increasing the run-off. Further, it encourages erosion of surface soil leading to nutrient removal in the running water. More ever, due to heavy clay content, the capillary porosity is relatively high resulting in impeded draina and reduced soil conditions. This results in increase of some soil elements to the level of toxicity to the plants. It also induced nutrient fixation in the clay complex thereby making the nutrient becoming unavailable to the crop, eventually causing deficiency of nutrients. Such soils are spread over Tamil Nadu in an area of 7,54,631 ha, which is 7.5% of total geographical area.

Management

The constraints in such soils can be managed by adopting suitable practices like

1. Provision of drainage facilities either through open or closed sub surface drains.
2. Forming contour and compartmental bunding to increase the infiltration rates of soils.
3. Application of huge quantities of river sand or red soils of coarser texture to dilute heaviness of the soil.
4. Application of liberal doses of organic manures like Farm Yard Manure, Compost, Green manure, Composted coir pith, sewage waste, press mud etc.
5. Adopting ridges and furrows, raised beds, broad bed and furrow systems.
6. Application of soil conditioners like H-concentrate, Vermiculite, Jalasakti etc to reduce run-off and soil erosion.

Excessively Permeable Soils

Excessively permeable soils are those having high amount of sand exceeding 70 per cent. Due to this, the soils are inert and unable to retain nutrient and water. These soils being devoid of finer particles and organic matter, the aggregates are weakly formed, the non-capillary pores dominating with very poor soil structure. Due to low retaining capacity of the soils, the fertilizer nutrients are also lost in the drainage water. These soils are spread over 24, 12, 086 ha in Tamil Nadu (23% of total geographic area).

Management

The excessively preamble soils can be managed by adopting the techniques given below.

1. Compacting the field with 400kg stone roller (tar drum filled with 400 kg of sand or stones can also be used) 8-10 times at optimum moisture conditions.
2. Application of clay soil up to a level 100 t ha^{-1} based on the severity of the problem and availability of clay materials.
3. Application of organic materials like farm yard manure, compost, press mud, sugar factory slurry, composted coir pith, sewage sludge etc.
4. Providing asphalt sheet, polythene sheets etc. below the soil surface to reduce the infiltration rate.
5. Crop rotation with green manure crops like Sunhemp, sesbania, daincha, kolinchi etc.

Sub soil hardening /hard pan

The sub soil hard pan in red soils in due to illuviation of clay to the sub soil horizon coupled with cementing action of oxides of Fe, Al and Calcium carbonate, which increases the soils bulk density to more than 1.8 Mg m^{-3}. Further, the hard pan can also develop due to continuous cultivation of crops using heavy develop due to continues cultivation of crops using heavy implements up to certain depth constantly. Besides, the higher exchangeable sodium content in black soils areas also result in compactness. All put together lowered the infiltration and percolation rates, nutrient movement and free air transport within the soils profile. It prevents root proliferation and limits the volume of soils available for nutrients uptake resulting in depleted, less fertile surface soil. Due to this, the contribution of sub soil fertility to crop growth is hampered. The area under this constraint is 10,54, 661 ha in Tamil Nadu (10: 48% TGA).

Management

These soils are managed by adopting following practices

1. Ploughing the soil with chisel plough at 0.5m interval criss cross at 0.5m depth once in 2-3 years.
2. Application of organics to improve the aggregation and soil structure so as to prevent further movement of clay to the lower layers.
3. Deep ploughing of the field during summer season to open up the sub soils.
4. Cultivating deep rooted crops like tapioca, cotton so as to encourage natural breaking of the hard pan.
5. Raising deep rooted semi perennial crops like Mulberry, Jasmine, Match wood tree etc. can also help in opening up the sub surface hard pan.

Surface crusting

Surface crusting is due to presence of colloidal oxides of iron and Aluminium in Alfisols which binds the soil particles under wet regimes. On drying it forms a hard mass on the surface. The ill effects of surface crusting are:

1. Prevents germination of seeds
2. Retards/inhibits root growth.
3. Results in poor infiltration.
4. Accelerates surface run off
5. Creates poor aeration in the rhizosphere
6. Affects nodules formation in leguminous crops Area : 4,51,584 ha (4.49% TGA) in Tamil Nadu.

Management

Surface crushing can be managed as below

1. When the soil is at optimum moisture regime, ploughing is to be given.
2. Lime at 2 t ha^{-1} may be uniformly spread and another ploughing given for blending of amendment with the surface soil.
3. Farm yard manure at 10 t ha^{-1} or composted coir pith at 12.5 t ha^{-1} or other organics may be applied to improve the physical properties of the soils, after preparation of land to optimum tilth.
4. Scraping surface soil by tooth harrow will be useful.
5. Bold grained seeds may be used for sowing on the crusted soils.
6. More number of seeds/trill may be adopted for small seeded crops.
7. Sprinkling water at periodical intervals may be done whenever possible.
8. Resistant crops like cowpea can be grown

Fluffy paddy soils

The traditional method of preparing the soil for transplanting rice consists of puddling, which substantially breaks soil aggregates into a uniform structure less mass. Under continuous flooding and submergence of soil for rice cultivation in a cropping sequence of rice-rice-rice, the soil particles and always in a state of flux and the mechanical strength is lost leading to the fluffiness of the soils. Impact of fluffiness is sinking of drought animals and labourers during pudding. This has been thus, an invisible drain of finance for the farmers due to high pulling power needed for the bullocks and slow movement of labourers during the puddling operations. Further fluffiness of the soil lead to very low bulk density and thereby leading to very rapid hydraulic conductivity and in turn the soil does not provide a good anchorage to the roots and the potential yield of crops is adversely affected. About 25, 919 ha (0.26% TGA) in Tamil Nadu have this problem.

Management

Following practices are needed to be adopted to overcome this problem.

1. The irrigation should be stopped 10 days before the harvest of rice crop.
2. After the harvest of rice, when the soil is under semi-dry condition, compact the field by passing 400 kg stone roller or an tar drum filled with 400 kg of sand for 8 times.
3. The usual preparatory cultivation is carried out after compaction.

Shallow soils

The shallow soils are characterized by the presence of the parent root immediately below the soil surface at about 15-20 cm depth. This restricts the root elongation and spreading. Hence, the crops grown in these soils necessarily are shallow noted crops, which can exhaust the soil within 2-3 seasons. Therefore, frequent renewal of soil fertility is a must in these soils. These soils can be managed by growing crops which can withstand the hard rooky sub soils like mango, ber, fig, country, goose berry, west, Indian cherry, Anona, Cashew, and Tamarind etc. These soils spread over an area of 1,16,509 ha in Tamil Nadu, which is 1-16 per cent of total geographical area.

Saline soils

Soils having higher proportion of soluble salts affect adversely the growth of plants. The salt level in saline soils exceeds a limit of 4.0 dSm^{-1}. Mostly these soils are dominant with chloride and sulphate salts. These salts are neutral salts and hence the pH of these soils may not round 8.5. Saline soils are formed through a soil forming process called salinization in semi arid and arid zones. Salinization refers to accumulation of soluble salts in the soil surface horizons.

Management of saline soils

1. Crop management

Growing crops that are tolerate high level soil salinity e.g.: Cotton, Ragi, Barley, sugar beat, Beet root, curry leaf, Bermuda grass, saline grass, spinach etc. Crops that are tolerant to soil salinity at medium level are paddy, wheat, onion, maize, sunflower, castor, grape, pomegranate, tomato, cabbage and potato. Crops that are tolerant to low level of soil salinity are garden beans, Reddish, lime etc., Blackgram, greengram are sensitive to soil salinity. Crops are to be chosen based on the soil salinity level.

2. Soil/cultural management

Growing crops in raised beds will reduce accumulation of salt a around root zone. Planting seedlings / sowing seeds on slopry ridges slecreases accumulation salts around root zone. Mulching soil prevents evaporation which reduces accumulation of salts due to capillary rise of water ate surface of soils. Providing drainage in water lugged areas also helps to reduce salt accumulation.

3. Fertilizer Management

Addition of extra dose of nitrogen to the tune of 20 – 25% of recommended level will compensate the low availability of N in these soils. Addition of organic manures like, FYM, compost, etc helps in reducing the ill effect of salinity due to release of organic acids produced during decomposition. Green manuring (Sunn hemp, daincha , Kolingi) and / or green leaf manuring also counteracts the effects of salinity.

4. Irrigation management

Proportional mixing of good quality (if available) water with salin water and then using for irrigation reduces effect of salinity. Alternate furrow irrigation favours growth of plant than flooding. Drip and sparkler irrigation systems aim are reduced use of water which is favorable for growth of plant since slat accumulation also reduced with low usage of water.

All the above four management practices suitably integrated to reduce the soil salinity which in favorable for better growth of plants and ultimately for better yields management of saline soils becomes essentials and unavoidable particularly in areas where both soil as well as irrigation water are saline in nature.

Sodic soils

Sodic soils are having high proportion of sodium at exchange complex. The sodium ion at exchange complex usually exceeds 15 percentages in this soils. These soils also have high proportion of carbonates and bicarbonates and hence the pH always more than 8.5. precipitated $CaCO_3$ present in this soil insoluble in nature. One contrary degraded sodic soils have low pH at surface but exchangeable sodium percentage is more than 15. These soils do not have precipitated $CaCO_3$. Sodic soils are formed due to the soil forming process of

alkalization (accumulation sodium in soils) which solodi solids / degraded alkali (sodic) formed by the process called solidization.

Reclamation of sodic soils

Physical

This is not actually removes sodium from exchange complex but improve physical condition of soil through improvement in infiltration and aeration.

a. Deep ploughing is adopted to break the hard pan developed at subsurface due to sodium and improving free-movement water. This also help in improvement of aeration.
b. Providing drainage is also practiced to improve aeration and to remove further accumulation of salts at not gone.
c. Sand filling which reduces heaviness of the soil and increases capillary movements of water.
d. Profile inversion – Inverting the soil benefits in improvement of physical condition of soil as that of deep ploughing.

Biological

Biological reclamation aims at improvement in physical condition of soils through addition of organic matter and to some extent solubilized native Ca for reclamation. Biological reclamation is also not a complete reclamation which not reduces sodium from soil below 15 percentages.

a. Addition of organic matter (Farm yard manure, press mud, Green manures, Green leaf manures) improves general physical condition of these soils. Further decomposition of organic matter releases organic acids and inorganic acids which are contracts the ill effects of high pH of soils.
b. Growing grasses like Cynodon dactylon, Brachiaria mulica, Chlon's gayana, pai korai etc which are become sodium from the soils. But the removal is slow process.
c. Afforestation i.e growing trees which alsoadd organic matter through leaf fall.

E.g. *Azadirachta* indica, *Prosopis juliflora*, *Tamarid*, *Albizia* sp., *Zizyphus* sp., *Acaciasp.*, *Cassia* sp. etc.

Chapter 14

Organic Agriculture and Climate Change

Climate change mitigation is urgent, and adaptation to climate change is crucial, particularly in agriculture, where food security is at stake. The gases contributing to the greenhouse effect mainly include carbon dioxide (CO_2), nitrous oxide (N_2O) and methane (CH_4). These gases have varying global warming potentials, which can be expressed in CO_2 equivalents.

Agriculture, currently responsible for 20-30% of global greenhouse gas emissions (counting direct and indirect agricultural emissions), can however contribute to both climate change mitigation and adaptation. The main mitigation potential lies in the capacity of agricultural soils to sequester CO_2 through building organic matter. This potential can be realized by employing sustainable agricultural practices, such as those commonly found within organic farming systems. Examples of these practices are the use of organic fertilizers and crop rotations including legume leys and cover crops. Mitigation is also achieved in organic agriculture through the avoidance of open biomass burning, and the avoidance of synthetic fertilizers, the production of which causes emissions from fossil fuel use.

Organic agriculture is associated with higher carbon sequestration as many organic practices help to improve soilquality and carbon sequestration. The most common organic practices that increase soil organic carbon are the use of organic fertilizers (such as the composted waste products from livestock husbandry), crop rotation involving legumes and the planting of cover crops.

Organic farming offers several ways to mitigate climate change when compared to conventional agriculture:

Primarily, organic farming, through its key practices of organic fertiliser use and crop rotations with forage legumes, tends to increase soil organic carbon levels resulting in carbon sequestration.This contributes to climate change mitigation, as it absorbs CO_2 from the atmosphere and stores the additional carbon in the soil. However, depending on soil type and climatic conditions, this process usually comes to a halt after some decades, when soil organic carbon levels have reached a new equilibrium and soils arethus saturated with respect to organic carboncontents. Furthermore, this storage of organiccarbon is reversible and the carbon can again be released into the atmosphere as carbon dioxide when switching to unsustainable practices.

Secondly, organic farming does not use mineral fertilisers. Thus, the emissions from industrial fertiliser production are avoided. In contrast to carbon sequestration, this is a permanent mitigation benefit that can be realised every yearanew.

Thirdly, organic farming generally has higher nitrogen use efficiencies and lower nitrogenuse levels than conventional agriculture. Thisresults in correspondingly lower emissions of the potent greenhouse gas nitrous oxide from fertilised soils, which is another straightforward and permanent mitigation benefit.

and **Finally**, organic farming tends to work with lower stocking densities of animals with respect to the land area available for grazing and feed production. These lower animal numbers go along with lower direct animal-related greenhouse gas emissions per farm, smaller manure quantities, and correspondingly reduced methane and nitrous oxide emissions from manure management.

The emission reduction potential of Organic Agriculture

Organic Agriculture can significantly reduce carbon dioxide emissions. As a viable alternative to shifting cultivation, it offers permanent cropping systems with sustained productivity. For intensive agricultural systems, it uses significantly less fossil fuel in comparison to conventional agriculture. This is mainly due to the following factors,

- Soil fertility is maintained mainly through farm internal inputs (organic manures, legume production, wide crop rotations etc.),
- Energy-demanding synthetic fertilizers and plant protection agents are rejected, and,
- External animal feeds - often with thousands of transportation miles - are limited to a low level.

As a consequence, the organic variants have in most cases a more favourable energy balance. Nevertheless there are reasons for organic farmers to do more to further reduce their dependency on fossil fuel and there are reasons to pay attention to the energy use on the food distribution system.

In avoiding methane, Organic Agriculture has an important though not always superior impact on reduction. Through the promotion of aerobic microorganisms and high biological activity in soils, the oxidation of methane can be increased. Secondly, changes in ruminant diet can reduce methane production considerably. However, technology research on methane reduction in paddy fields – an important source of methane production - is still in its infancy.

Nitrous oxides are mainly due to overdoses and losses on nitrogen. These are effectively minimized in Organic Agriculture because:

- No synthetic nitrogen fertilizer is used, which clearly limits the total nitrogen amount and reduces emissions caused during the energy demanding process of fertilizer synthesis.
- Agricultural production in tight nutrient cycles aims to minimize losses.
- Animal stocking rates are limited. These are linked to the available land area and thus excessive production and application of animal manure is avoided.
- Dairy diets are lower in protein and higher in fibre, resulting in lower emission values.

Using biomass as a substitute for fossil fuel represents another emission reduction option. Organic Agriculture is well positioned in this sector. It has the advantage that, inorganic N-fertilizers are not applied, which cause significant emissions of N_2O and use a lot of energy.

The sequestration potential of Organic Agriculture

Organic Agriculture has a particular sequestration potential as it follows the key principle of tight nutrient and energy cycles through organic matter management in soils. This is achieved through improved practices in cropland management and in agroforestry.

Various long-term trials provide evidence that the regular addition of organic materials to the soil is the only way to maintain or even increase soil organic carbon (SOC). The systematic development and application of organic fertilization technologies has been the domain of Organic Agriculture for many decades and outstanding results have been achieved so far. Key issues of technology development have been:

- To optimise the quantity and application of organic manure. A close integration of crop production and animal husbandry and the systematic recycling of organic waste are basic elements.
- To improve organic waste processing techniques to obtain high quality manure. Through composting of animal and plant residues losses in the humification process are minimized and a higher proportion of the solid humus fraction is achieved.

Long and diversified crop rotations and legume cropping are further characteristics of Organic Agriculture that help to increase SOC.

In conventional agriculture, conservation tillage is largely promoted as a measure to sequester carbon dioxide. This technology combines minimum tillage with organic covers, herbicides and often herbicide resistant GMO crops. Both of the last two are prohibited in Organic Agriculture. Latest research results revealed that gains in soil organic carbon have been overestimated and are partly or completely offset by increased N_2O emissions. Thus it can be concluded that minimum tillage combined with mineral fertilizer application compares less well with Organic Agriculture if the focus is on GHGs in general rather than considering carbon sequestration alone. The task of Organic Agriculture will be to integrate conservation tillage in a way that negative effects are avoided.

Agroforesty – a management system that integrates trees in the agricultural landscape – is another technology that is systematically applied in Organic Agriculture. It is a feasible method to succeed shifting cultivation systems but also to improve and add value to low productive cropland. Agroforestry holds the biggest potential of agricultural carbon sequestration in tropical countries.

It is worth noting that the sequestration of carbon, i.e., an increase of soil organic matter is also leading to more fertile soils, better water retention capacity and reduced nutrient leakage.

Mitigation

Mitigation is primarily achieved through long established and optimized organic farming practices. These prac-tices include enhancing soil biological processes, soil fertility and structure, creating organic matter in forms that are more effective at producing soil carbon, integrating crop and livestock systems, increasing the propor-tion of vegetation cover which promotes the soil's micro-organisms that stabilize soil carbon and through en-couraging and facilitating local production and consumption. The more widespread adoption of organic farm-ing practices and grass-based and mixed farming systems can make a significant contribution to greenhouse gas mitigation.

1. Maintaining and increasing soil organic carbon in agricultural systems is the option with the largest mitigation option in agriculture. Organic agriculture has a significant potential contribution in this respect: practices that are commonly used on organic farms (use of organic fertilizers, fertility building leys with legumes and cover crops) further the production of soil organic matter.
2. Organic agriculture has lower N_2O emissions from nitrogen application, due to lower overall nitrogen input per ha than in conventional agriculture
3. Open burning of crop residues and biomass waste is prohibited for agriculture in most industrialized countries, but it is still common practice in conventional agriculture in many developing countries. In organic agriculture, biomass is not burned, but recycled to the soil to improve fertility. This reduces

the CH_4 and N_2O emissions in comparison to conventional agriculture, where crop residues are often burnt on the field

4. Optimal manure management: Improved manure management including distribution systems, such as slurry injections into soils or drag hoses, reduce nutrients losses considerably. Covering manure and slurry storage sites reduces CH_4 emissions. In addition, CH_4 can be captured and used as biogas. Specifically in organic agriculture, manure is often composted. In aerobic composting, assuring sufficient aeration will avoid CH_4 and N_2O emissions. However, decomposition of organic material leads to emissions if it depletes the available oxygen hence the need to optimize aeration in aerobic composting systems. Partial microbial digestion of farmyard manure such as through composting for example promotes its potential to be converted into securer forms of soil carbon.
5. Conventional stockless arable farms depend on the input of synthetic nitrogen fertilizers, while stockpiled manure and slurry on livestock farms create additional emissions and other environmental problems. Organic farms mitigate such problems by on-farm or cooperative use of farmyard manure between crop and livestock operations. In particular where this leads to an overall increase in N useefficiency, the result is a reduction in emissions per kg of product.
6. Combining perennial and annual crops: Organic agriculture systems combine different perennial plant species such as trees, shrubs, palms and grasses with different annual crops. Organic agriculture also integrates higher levels of native perennial trees and hedgerows from the surrounding landscape. Plants sequester CO_2 from the atmosphere by pulling it in via photosynthesis. Perennial plants develop their roots and branches over many years to store carbon in the vegetation and soil, annual plants however leave no permanent vegetation and comparatively little soil carbon.
7. There is great potential to improve mitigation by increasing the use of perennials and reducing annual crops, especially those grown to provide concentrated feed for industrial

livestock production. In agro-forestry systems where trees are combined with crop production the potential is even higher for sequestrating carbon. Diverse agro-forestry systems have mostly been developed for the humid tropics but a combination of annual and perennial crops could significantly raise carbon sequestration and food and fuel production in temperate regions. The inclusion of permanent species in tropical farming systems can significantly improve productivity and can encourage farmers to avoid traditional shifting cultivation or slash and burn agriculture and the associated GHG emissions. Combining permanent and annual crop species also enhances the eco-functionality and productivity of the farming system.

8. Due to reduced concentrate feed use in ruminant animal husbandry, organic animal agriculture causes less direct land use change (deforestation to gain cropland for concentrate feed production) and thus also less CO_2 emission from soil carbon losses due to this change. Since organic grass-land and fodder production is often equally productive as with conventional systems there are little in-direct land use change effects. On the other hand, higher roughage diets can lead to higher methane emissions from ruminants. The net effect of increased roughage nonetheless yields an overall reduction in emissions. Research comparing ruminant livestock production systems also shows that organic farming systems perform favourably, in terms of energy use, due to energy savings associated with reliance on clover-grass leys and high forage/low cereal diets. In addition, animal welfare is improved, as a high roughage diet is more natural for ruminants. Furthermore in-creased longevity within organic systems reduces the relative emissions from the unproductive rearing phase of dairy cows.

9. Further mitigation options in agriculture include a) on-farm biogas production from agricultural waste, in particular from manures and crop residues; b) optimized manure and slurry management (optimized stables) and storage (coverage); c) use of resistant varieties and effective crop breeding (to reduce energy needs for spraying and to increase productivity); d) reduced tillage to increase soil carbon contents; e) increased

efficiency of machinery and their use (e.g. optimal tire pressure and speed on the fields) and of buildings to reduce energy use; f) non-permanently flooded rice production; this avoids methane emissions, but may increase N_2O emissions, in particular at high N rates.

10. Avoiding bare soil: Ensuring the soil is always covered with vegetation prevents the soil from being exposed to processes that accelerate GHG emissions from stored soil carbon. Avoiding bare fallows through the inclusion of catch crops and green manures within organic farming systems retains nutrients for future utilization and avoids the emissions associated with additional nitrogen inputs. Cover vegetation is also important in providing a greater and more continuous supply of the root exudates that support the soil's micro-organisms which build the soil carbon store. Crop residues can also be left on the field as a protective layer for the soil and avoid CH_4 and N_2O emissions caused by the burning of crop residues in conventional agriculture.

Adaptation

Agriculture is highly vulnerable to climate change and our food supply relies on successful adaptation. Adapta-tion actions include those necessary to restore the resilience of eco-systems and their productivity to enable sustainable economic development.The financial requirements of organic agriculture for adaptation are low. Additional costs mainly come from information provision, education and extension services. Organic agriculture however offers innovative farmer based group systems that facilitate best practice knowledge exchange in a systematic and cost free manner. This is particularly important for the empowerment of vulnerable and poor people in rural populations that rely on agriculture for their livelihoods.

1. Farm practices commonly used within organic agriculture increase and stabilize **soil organic matter**. As a result, soils under organic management can better capture and store water than soils of conventional cultivation. Organic production is thus less prone than conventional cultivation to extreme weather

conditions, such as drought, flooding, and water-logging, which are expected to become much more frequent under climate change. Organic farming practices have also been shown to reduce soil erosion, increase aggregate stability and stimulate soil biological activity. Organic agriculture thus provides protective responses to key consequences of climate change, particularly those associated with increased occurrence of extreme weather events, storms, droughts and floods. As already mentioned, these benefits do not depend on the implementation of OA as a whole system, but on implementation of certain key practices such as recycling of manures and crop residues through organic fertilizers, which can also be implemented in conventional agriculture.

2. Organic agriculture also increases soil quality and fertility, with regard to soil nutrients, improved soil structure and aeration, water retention capacity and thus water availability. The biological diversity of soil microbes, insects and earthworms is increased, all of which have important roles for soil quality.Organic agriculture uses a greater level of diversity among crops, crop rotations and production practices than commonly employed in conventional, industrialized agriculture, which often is based on monocultures. Organic farms have a generally **higher biodiversity**, also due to set-aside areas and other landscape elements. This improves ecological and economic stability. The diversity of income sources, as well as the resilience to adverse effects of climate change is thus increased. An example of the benefits is the enhanced biodiversity, which reduces pest outbreaks and severity of plant and animal diseases, while also improving utilization of soil nutrients and water. For improving resilience to a higher occurrence of heat waves under climate change, the use of agro-forestry and shade trees can be a very efficient mechanism for lowering critical temperatures. The-se diverse systems may also enhance carbon sequestration.
3. Organic agriculture is a low-risk farming strategy based on lowering external inputs and optimizing biological functions. Besides lowering toxicity, reduced inputs lower costs and thus

con-tribute to the competitiveness of organic agriculture economically. In addition, organic price premiums may be realized. These factors working together can lower the financial risks and improve the rewards. They provide a type of low cost but effective insurance against crop reduction or failure. Due to this increased coping capacity of the farms, the risk of indebtedness in general is lowered. Organic agriculture is thus most often a viable alternative for poor farmers. Risk management, risk-reduction strategies, and economic diversification to build resilience are also prominent aspects of adaptation to climate change.

4. Agro-genetic biodiversity: Organic agriculture has a big role to play in protecting the world's agricultural genetic resources (e.g. crop and farm animal species and varieties). Organic agriculture encourages the use of locally adapted varieties and decentralized participatory breeding programs especially in-situ (on-farm) based conservation, breeding and production. In-situ approaches maintain varieties for future needs while allowing them to continuously adapt to environmental pressures such as climate change. Our agro-genetic diversity is a critical resource in the effective ongoing adaptation to continuous changes in climate.
5. Diversification: Diversification is a fundamental aspect of organic agriculture. Resiliency to climate disasters is closely linked to farm biodiversity. Practices that enhance biodiversity allow farms to mimic natural ecological processes, enabling them to better respond to change and reduce risk. Biologically diverse organic farms that optimize ecological functionality avoid the build-up of disease and pest levels and are more resilient to other environmental pressures. Crop diversity (both temporal and spatial) provides a variety of rooting depths that enhance soil stability and structure, improves nutrient and water use, and con-tributes to a stabilized microclimate. Farmers who increase inter-specific diversity via organic agriculture suf-fer less damage compared to conventional farmers planting monocultures. The diversity of landscapes, farming activities and crops is greatly enhanced in organic agriculture resulting

in farming systems that are re-silient to the adverse effects of climate change. The robust and resilient nature of organic agriculture systems also helps to protect sequestrated carbon from climatic disturbances increasing the permanence of sequestra-tion.

6. Resilient crops: Building healthy soils is the basis for growing healthy resilient plants that are better able to withstand environmental pressures such as increased water scarcity and pest and disease pressure. Organic agriculture strengthens the immune systems of plants and thus the defence and self-healing capacities of crops against pests and diseases. Plants that obtain their nutrients through natural biological processes are more resilient to environmental stress than crops that obtain their primary nutrition artificially through highly soluble chemical fertilizers. This is mainly achieved through optimal soil and water management, the building of soil structure and fertility and the choice of locally adapted robust crop varieties. In addition organic crops tend to have longer and denser roots that are able to seek out water reserves deeper in the soil profile and which are also more resilient to desiccation.

7. Organic agriculture provides a good opportunity to utilize local and indigenous farmer knowledge, adaptive learning and crop development, which are seen as important sources for adaptation to climate change and variability in farming communities. However, it is important to stress that existing local knowledge, in the front of climate change, needs to be updated by more intensive observations and their interpretation, as well as with the assistance of research, experimentation and innovation.

8. Preventing and reversing soil erosion and restoring degraded land: Organic practices that contribute to the mitigation of global warming also increase soil health: use of organic instead of chemical fertilizer; cover crops, catch crops, green manure; composting; appropriate tillage; and the integration of perennials and trees into the farming system. Healthy soils have higher organic matter contents and greater biological activity

which improves soil structure and stability. The organic management of soils is a highly effective tool to regenerate degraded land. By increasing the soil organic matter content, organic ag-riculture is able to build up soil carbon and soil fertility from low levels and bring severely degraded land back into production.

9. Further adaptation options include plant breeding for improved drought and heat resistance, use of locally adapted varieties (e.g. some traditionally grown local varieties) and optimized feeding practices to avoid heat stress for animals (e.g. early morning or night pastures).

Annexures

Annexure 1: Certification Agencies In India

IMO Control Pvt. Ltd.
Mr. Umesh Chandrasekhar
Director
No. 1314, Double Road
Indiranagar 2nd Stage
Bangalore-560 038. (Karnataka)
Tel.: 080-25285883, 25201546Fax: 080-25272185
Email:imoind@vsnl.com

Bureau Veritas Certification India Pvt.Ltd.
(Formerly Known as BVQI (India) Pvt. Ltd.)
Mr. R. K. Sharma Director
Marwah Centre, 6th Floor
Opp. AnsaIndustrial Estate
Krishanlal Marwah Marg Off Saki-Vihar Road
Andheri (East) Mumbai-400 072 (Maharashtra)
Tel.: 022-56956300, 56956311
Fax No. 022-56956302 / 10
Email: scsinfo@in.bureauveritas.com

Indian Organic Certification Agency (INDOCERT)
Mr. Mathew Sebastian
Executive Director
Thottumugham P.O.Aluva-683 105
Cochin, (Kerala)
Tel.: 0484-2630908-09/2620943
Email:Mathew.Sebastian@indocert.org

ECOCERT India Pvt. Ltd
Dr. Selvam Daniel (C.R.)
Sector-3, S-6/3 & 4, Gut No. 102
Hindustan Awas Ltd. Walmi-Waluj Road
Nakshatrawadi Aurangabad – 431 002 (Maharashtra)
Tel.:: 0240-2377120, 2376949
Fax No.: 0240-2376866
Email: ecocert@sancharnet.in

Lacon Quality Certification Pvt. Ltd
Mr. Bobby Issac
Director
Chenathra, Theepany, Thiruvalla - 689101. (Kerala)
Telefax: 0469 2606447
Email: laconindia@sancharnet.in

Natural Organic Certification Agency
Mr. Sanjay Deshmukh, CEO
Chhatrapati House Ground Floor
Near P. N. Gadgil Showroom
Pune-411 038 (Maharashtra)
Tel.: 020-25457869, 56218063Fax: 020-2539-0096
Email: contact@nocaindia.com

SGS India Pvt. Ltd.
Dr Manish Pande
Divisional Manager – Food,
Retail& CSRS250 Udyog Vihar Phase – IV
Gurgaon – 122 015 (Haryana)
Tel.: 0124-2399990-98, Fax No.: 0124-2399764
Email: namit_mutreja@sgs.com

OneCert Asia Agri Certification Pvt. Ltd.
Mr. Sandeep Bhargava
Chief Executive Officer
Agrasen Farm, Vatika Road,
Vatika P.O., Off Tonk,
Jaipur-303 905, (Rajasthan)
Tel.: 0141-2770342, Telefax No: - 0141-2771101
Email: info@onecertasia.in

Control Union Certifications
(Formerlyknown as Skal International (India)
Mr. Dirk Teichert
Managing Director"Summer
Ville"8th Floor 33rd – 14th Road
Junction Off Linking Road, Khar (West)
Mumbai –400052 (Maharasthra)
Tel.: 022-67255396/97/98/99Fax 022-67255394/95
Email: cuc@controlunion.in
cucindia@controlunion.com, controlunion@vsnl.com

Uttranchal State Organic CertificationAgency (USOCA)
Director 12/II Vasant Vihar
Dehradun-248 006 (Uttaranchal)
Tel.: 0135-2760861Fax: 0135-2760734
Email: uss_opca@rediffmail.com

APOF Organic Certification Agency (AOCA)
Mr. Swapnil Satish
General Secretary
Holkar House, First Floor, Sr No: 54
Near Nikhil Garden,
Wadgaon Bk, Pune 411041
Phone /fax: 020-65410070, Mob.: +91 7720073202
Email: info@aoca.in

Food Cert India Pvt. Ltd
Quality House, H. No. 8- 2- 601/P/6
Road No. 10, Banjara Hills,
Panchavati Colony, Hyderabad – 500 034
Tel.: +91- 40-23301618, 23301554, 23301582 Fax: +91-40-23301583
Email: foodcert@foodcert.in

Rajasthan Organic CertificationAgency (ROCA)
3rd Floor, Pant Krishi Bhawan, Janpath
Jaipur 302 005(Rajasthan)
Tel.:: 0141-2227104, Tele Fax: 0141-2227456
Email: dir_rssopca@rediffmail.com

Aditi Organic Certifications Pvt. Ltd
No. 531/A, Priya Chambers
Dr. Rajkumar Road, Rajajinagar,1st Block
Bangalore - 560010
Tel.: +91-80-32537879Fax: +91-80-23373083 Mobile: +91-9845064286
Email: aditiorganic@gmail.comWebsite: www.aditicert,net

Annexure 2: Traditional Crop Varieties of Southern India

Rice varieties

1. Kullakkar
2. Annamazhagi
3. Arubhathanguruvai
4. Poongaar
5. Kerala ragam
6. Kuzhiyadichaan
7. Kullangaar
8. Mysore malli
9. Kudavaazhai
10. Kaattuyaanam
11. Kaattupponni
12. Vellaikkaar
13. Manjal ponni
14. Karuppu seeragasamba
15. Katti samba
16. Kuruvikkaar
17. Varappu kudainchaan
18. Kuruvai kalanjiam
19. Kambanjsamba
20. Bommi
21. Kaalaa namak
22. Thiruppathisaaram
23. Pisini
24. Vellai kuruvikkar
25. Vishnubogam
26. Mozhikkaruppu samba
27. Kaattu samba
28. Karunguruvai
29. Thengaaipoo samba
30. Kaattu kuthalam
31. Basumathi
32. Puzhuthi samba
33. Paal kudavaazhai
34. Vaasanai seeraga samba
35. Kosuva kuththalai
36. Iluppaipoo samba
37. Thulasivaasa seeraga samba
38. Chinnapponni
39. Vellaipponni
40. Sigappu ponni
41. Sigappu kavuni
42. Kottaara samba
43. Seeraga samba
44. Kaivirach samba
45. Kandhasaalaa
46. Panangaattu kudavaazhai
47. Sanna samba
48. Iravai paandi
49. Sembili samba
50. Navara
51. Karuthakkaar
52. Kichili samba
53. Kaivara samba
54. Salem sanna
55. Thooya malli
56. Vaazhaipoo samba
57. Arcadu kichili
58. Thanga samba
59. Neela samba
60. Manalvaari
61. Garudan samba
62. Kattai samba

63. Aathur kichili
64. Kundhaavi
65. Sigappu kuruvikkaar
66. Koombaalai
67. Vallaragan
68. Kavuni
69. Poovan samba
70. Muttrina sannam
71. Chandikaar
72. Karuppu kavuni
73. Mappillai samba
74. Madumuzhungi
75. Ottadam
76. Vaadan samba
77. Samba mosanam
78. Kandavaari samba
79. Vellai milagu samba
80. Kaadai kazhuththaan
81. Neelanj samba
82. Javvathumali nel
83. Vaigunda
84. Kappakkaar
85. Kalian samba
86. Aduku nel
87. Sengaar
88. Rajamannaar
89. Murugan kaar
90. Sornavaari
91. Soorakkuruvai
92. Vellai kudavaazhai
93. Soolakkunuvai
94. Norungan
95. Perungaar
96. Poompaalai
97. Vaalaan
98. Kothamalli samba
99. Sorna masoori
100. Bayakundaa
101. Pachai perumaal
102. Vasara mundaan
103. Konakkuruvai
104. Puzhuthikkaar
105. Karuppu basuamthi
106. Veedhi vadangaan
107. Kandasaali
108. Amyo mogar
109. Kollikkaar
110. Rajabogam
111. Sembini ponni
112. Perum koombaalai
113. Delli pogalu
114. Kachcha koombaalai
115. Mathimuni
116. Kallurundaiyaan
117. Rasagadam
118. Kambam samba
119. Kochin samba
120. Sembaalai
121. Veliyaan
122. Raajamudi
123. Arubhathaam samba
124. Kaattu vaanibam
125. Sadaikkaar
126. Samya
127. Maranel
128. Kallundai
129. Sembini priyan
130. Kashmir dal

131. Kaar nel
132. Mottakkoor
133. Raamakalli
134. Jeera
135. Sudarhaal
136. Padhariya
137. Sudhar
138. Thimari kamodu
139. Jaljera
140. Mal kamodu
141. Ratnasudi
142. Haalu uppalu
143. Siddha sanna
144. Varedappana sen
145. Sittigaa nel
146. Karigajavali
147. Karijaadi
148. Sannakki nel
149. Katka
150. Singini kaar
151. Sembaalai
152. Milagi
153. Vaal sivappu
154. Sivappu chitraikaar

Millet varieties

1. Karikattai Ragi
2. Kattucumbu – Pearl Millet
3. Thattai Varagu
4. Puvadan Keazhvaragu – Ragi
5. Sorghum (Kona Cholam)
6. Chittan samai
7. Sen Cholam (Wild Sorghum)
8. Kakacholam (Traditional Cholam)

Oil seed varieties

1. Karuppu ellu
2. Sivappu ellu
3. Kothu Kadalai – Groundnut Bunch variety
4. Kadalikai – Traditional Groundnut

Guava varieties

1. Nattu vellai (guava)
2. Senkoiyya

Jackfruit varieties

1. Parukkan pala
2. Varikkai chakkai pala
3. Koozhan chakkai pala
4. Mulagumoodu

Banana Varieties

1. Palayankodan
2. Nendran
3. Sevvazhai
4. Matti

MangoVarieties

1. Ram Seetha (*Annona reticulate*) – Local Anona Variety
2. Chengavarkai (Mango –Red Shaded)
3. Chetti ottu
4. Guava – Nattu Vellai (White) and Senkoiya (Red Flesh)

Vegetable Varieties

1. Kulithakali (Tomato)
2. Vazuthaalangai *(Solanum melongena)* – Egg Plant / Brinjal
3. Kannadi Katthari – Wild Brinjal

4. Poyyur Kathari – Local Brinjal Variety
5. Ramnad Gundu Chillies
6. Sivappu Vendai
7. Gandhari chilli
8. Neela milahai
9. Ther Kothavari
10. Karuppu Avarai
11. Nattu Pusani
12. Jaffna Moringa

Tuber crop varieties

1. Vellai bontha (Tapioca)
2. Karuppu bontha (Tapioca)
3. Kariyilai poriyan (Tapioca)
4. Adukku muttan (Tapioca)
5. Nattu seeni kizhangu (Yams)

Spices varieties

1. Nattu Milagu
2. 'Karimundan - Milagu
3. Kattumilagu
4. ''Valukkai' - cardamom
5. Malaippoodu - Garlic
6. Periumpoodu - Garlic
7. Kudampuli Camboge

Annexure 3: Traditional Livestock Breeds

S.No	Livestock breeds
1.	Athakkaruppan
2.	Azhukkumaraiyan
3.	Anarikaalan
4.	Aalaiverichchaan
5.	Aanaichchoriyan
6.	Kattaikkaalai
7.	Karumaraiyaan
8.	Kattaikkaari
9.	Kattukkomban
10.	Kattaivaal koolai
11.	Karumaraikkaalai
12.	Kannan mayilai
13.	Kathikkomban
14.	Kallakkaadan
15.	Kallakkaalai
16.	Kattaikkomban

S.No	Livestock breeds
17.	Karungoozhai
18.	Kazharvaaiveriyan
19.	Kazharchikkannan
20.	Karuppan
21.	Kaarikkaalai
22.	Kaarsilamban
23.	Kaarampasu
24.	Kuttaiseviayan
25.	Gundukkannan
26.	Kuttai naramban
27.	Kuthukkulamban
28.	Kullasivappan
29.	Koozhaivaalan
30.	Koodukomban
31.	Koozhaisivalai
32.	Kottaippaakkan

S.No	Livestock breeds	S.No	Livestock breeds
33.	Kondaiththalaiyan	63.	Soriyan
34.	Yerichuzhiyan	64.	Thalappan
35.	Yeruvaalan	65.	Thallayan kaalai
36.	Naaraikkazhuthan	66.	Tharikomban
37.	Nettaikomban	67.	Thudaiserkoozhai
38.	Nettaikkaalan	68.	Thoongachchezhiyan
39.	Padappu pidungi	69.	Vattappullai
40.	Padalai komban	70.	Vattachcheviyan
41.	Pattikkaalai	71.	Valaikkomban
42.	Panangai mayilai	72.	Vallikkomban
43.	Pasungazhuthaan	73.	Varnakkaalai
44.	Paalvellai	74.	Vattakkariyan
45.	Pottaikkannan	75.	Vellaikkaalai
46.	Ponguvaayan	76.	Vellaikkudumban
47.	Porukkaalai	77.	Vellaikkannan
48.	Mattaik kolamban	78.	Vellaipporaan
49.	Manjal vaalan	79.	Mayilai kaalai
50.	Maraichchivalai	80.	Vellai
51.	Manjali vaalan	81.	Kazhuthikapillai
52.	Manja mayilai	82.	Karukkamayilai
53.	Mayilai	83.	Panangaari
54.	Megavannan	84.	Santhanappillai
55.	Murikkomban	85.	Sarchchi
56.	Muttikkaalan	86.	Sindhumaadu
57.	Murikaalai	87.	Semboothukkaari
58.	Sannguvannan	88.	Sevalamaadu
59.	Semmaraikkaalai	89.	Naattumaadu
60.	Sevalai eruthu	90.	Erumaimaadu
61.	Semmaraiyan	91.	Kaarimaadu
62.	Senthaazhaivayiran		

Index

A

B

F

G

H

I

J

K

L

M

Q

R

S

T

U

V

W

Y

Printed in the United States
by Baker & Taylor Publisher Services